D1387962

VIROLOGY MONOGRAPHS

DIE VIRUSFORSCHUNG IN EINZELDARSTELLUNGEN

CONTINUATION OF / FORTFÜHRUNG VON

HANDBOOK OF VIRUS RESEARCH
HANDBUCH DER VIRUSFORSCHUNG
FOUNDED BY / BEGRÜNDET VON

R. DOERR

EDITED BY / HERAUSGEGEBEN VON

C. HALLAUER

17

SPRINGER-VERLAG

WIEN   NEW YORK

THE NATURE AND ORGANIZATION OF RETROVIRAL GENES IN ANIMAL CELLS

BY

D. R. STRAYER AND D. H. GILLESPIE

SPRINGER-VERLAG

WIEN   NEW YORK

DAVID R. STRAYER and DAVID H. GILLESPIE

Orlowitz Cancer Institute
Hahnemann Medical College and Hospital
Philadelphia, Pennsylvania, U.S.A.

QR
395
.S76

With 50 Figures

Library of Congress Cataloging in Publication Data. Strayer, David R. The nature and organization of retroviral genes in animal cells. (Virology monographs; 17.) Bibliography: p. 1. Viruses, RNA. 2. Genetic recombination. 3. Viral genetics. 4. Cell transformation. 5. Oncogenic viruses. I. Gillespie, David H., 1940— joint author. II. Title. III. Series. IV. Title: Retroviral genes in animal cells. QR 360. V 52. no. 17 [QR 395]. 576'. 64 s [616.99'2'0194]. 80—13727

ISSN 0083-6591
ISBN 3-211-81563-5 Springer-Verlag Wien-New York
ISBN 0-387-81563-5 Springer-Verlag New York-Wien

Contents

I. Introduction 1

II. Origin of Retroviruses 11

III. Organization of Endogenous Retrovirus Genes 21

IV. Organization of Infectious Retrovirus Genes 45

V. Horizontal Transmission of Retroviruses Among Animals 59

VI. Relatedness Among Retroviruses: Recombinant Viruses 69

VII. Human Retroviruses 85

    1. The Logical Problem 85

    2. The Natural History of Primate, Type-C, RNA Viruses 86

    3. Candidate Human Retroviruses 91

VIII. Retrovirus Integration, Growth Regulation and Human Cancer . . . 96

References 101

I. Introduction

RNA tumor viruses[1] have become increasingly utilized in studies of cellular transformation and gene regulation. The genes of retroviruses exist in two forms; as extrachromosomal, RNA-containing, infectious particles and as DNA proviruses[2] stably associated with cell genes. Components from the extracellular form can be collected in large quantity and purified for the preparation of molecular probes. These probes can be used to dissect the sequence of events required for the establishment and expression of the integrated form. Furthermore the genomes of retroviruses originated from normal cell genes, genes called virogenes[2]. The nucleic acid and protein probes isolated from these viruses are therefore useful for studying the nature and expression of this normal cell gene and in elucidating the physiological role of its products. RNA tumor viruses perhaps offer us one of the most complete sets of biochemical reagents and biological responses for examining gene regulation in vertebrates and for studying the consequences of aberrant gene regulation on cell growth in tissue culture and in animals. Furthermore, there is an increasing conviction that virogenes play an important role in normal development and/or differentiation (RISSER, STOCKERT and OLD, 1978). Consequently, there is a growing feeling that DNA proviruses are altered virogenes and are capable of interfering with normal development or differentiation, causing reprogrammed growth or the incapacity to specialize.

Before retroviruses can be most effectively used as experimental tools for studying the molecular basis of transformation and genetic regulation it will be necessary to understand several aspects of the molecular biology of the viruses

---

[1] A retrovirus is defined here as a replicating, enveloped particle about 1000 Å in diameter having an inner structure, the core or nucleoid, encircling high molecular weight RNA. The virion RNA is poly(A)-containing and has an associated RNA primer such that the complex is suitable for copying into DNA using virion RNA-dependent DNA polymerase (reverse transcriptase). The definition also extends to variants of this type of virus lacking certain components.

Retroviruses can be subdivided into types A, B, C and D by morphological criteria. All sarcoma and leukemia viruses and the endogenous viruses they evolve from are type C. Types A, B and D will not be considered in any detail.

Those retroviruses capable of causing cancer or suspected to have oncogenic potential will be called RNA tumor viruses or oncornaviruses.

[2] The provirus is the double-stranded DNA copy of the viral genes. The provirus can exist in a closed circular unintegrated form or can be inserted into a cell chromosome following infection by a retrovirus. The virogene is a gene(s) in a normal cell carrying viral or viral-related sequences and thought to be the progenitor of the RNA tumor virus genome.

themselves. The life cycle of RNA tumor viruses in the laboratory (TEMIN, 1974; GILLESPIE, SAXINGER and GALLO, 1975; TODARO and GALLO, 1976) and the bio-chemistry of the viral RNA-dependent DNA polymerase (SARIN and GALLO, 1974; TEMIN and BALTIMORE, 1972; BISHOP, 1978) have been reviewed by others. Several scientists have incorporated the known features of retrovirus molecular biology into a cell biology context, hypothesizing on such diverse topics as morpho-logical transformation (TEMIN, 1971; HUEBNER and TODARO, 1969), RNA processing (GILLESPIE, SAXINGER and GALLO, 1975; GILLESPIE and GALLO, 1975), immunogenesis (TEMIN, 1974), embryogenesis (TEMIN, 1971; HUEBNER and TODARO, 1969; GILLESPIE and GALLO, 1975) and natural selection (BENVENISTE and TODARO, 1974; GILLESPIE, MARSHALL and GALLO, 1972).

In this monograph we have tried to consider several features of the molecular biology of RNA tumor viruses that will be important when evaluating these and other proposals. We focused on information transfer, especially on the interaction between the invading virus and the cell chromosome. We have attempted to bring the concepts and results of previous years into line with some of the recent findings in molecular biology. In doing so we have imprinted some aspects of the overall picture with our own bias and with prevailing scientific opinion, but we have tried to indicate when this was the case.

In this connection we summarize below our working hypothesis on the nature of the interaction between an invading virus and its potential host cell, from the viewpoint of the DNA recombination event. This event, termed the "integration" of the virus genes into the chromosome of the infected cell appears to be the critical event with respect to whether the infected cell will become transformed, whether it will produce infectious virus, whether it will liberate virus-like but inactive particles, etc. The synthesis of DNA copies of parental viral RNA by reverse transcriptase and the integration of the "free proviral" DNA into host chromosomes, as TEMIN proposed in the early 1960's, is now a well docu-mented and generally accepted process. In the simplest view, one molecule of free proviral DNA interacts with DNA of the host chromosome and results in the insertion of a single, complete DNA provirus per genome of the infected cell. Much of the evidence suggesting such a simple model stems from highly artificial, often abnormal host cell/virus combinations, for example, cells infected by a virus they never see in nature. Subsequently, very special types of infected cells are often selected, for example, virus-producing cells, and they are cloned and recloned before their properties are catalogued.

The natural situation is likely to be considerably more complex. Some more or less recent findings require us to modify the basic scheme outlined in the paragraph above in order to make the events that occur during integration more compatible with the extremely complex nature of these viruses. Three of the recent findings are introduced below without supporting references; they will be discussed in detail in ensuing sections.

First, there exist multiple sites of provirus integration in the host chromosome (*int* sites), since some new viral sequences in a virus-infected cell are multiple-copy. The number of copies of a virus gene inserted during infection is not readily predictable. Infection of mammalian cells by avian viruses and *vice versa* would be expected to result predominantly in a single insertion of all or part of the viral genome, because of lack of homology between the virus genome and its *int* site on the host chromosome. Infection of cells by viruses that naturally interact with them would more often result in multiple insertions.

In some instances *int* sites may be endogenous virogene sequences (SHOYAB, DASTOOR and BALUDA, 1976) but newer results suggest that newly evolved repeated DNA may serve as integration sites for retroviruses as they do for the DNA virus, SV40 (LAVI and WINOCOUR, 1972; ROSENBERG *et al.*, 1977). These repeated *int* sites could also be recombinational "hot spots", like those GALLO (1974) proposed in an attempt to relate RNA tumor viruses to human leukemia.

Second, any given integration event can result in the insertion of whole or partial proviral fragments into the host chromosome and both types of integration might take place in the same cell. Thus, different portions of the viral genome could be inserted with different frequencies into a single host cell. The frequency and extent of integration depend on the titer of free provirus, the mechanism of provirus synthesis, the degree of homology between provirus and *int*, the number of *int* sites and probably other factors. Generally, the study of virus-infected cells is biased toward examinations of complete provirus because, ordinarily, virus-producing cells are selected.

Third, viruses can acquire cell genes, especially when only part of the provirus is inserted into the cell chromosome and when *int* is a virogene. Viruses released from infected cells are often different from the infecting virus(es). During genetic change the viruses can acquire new biological properties, such as sarcomagenic or leukemogenic potential.

Probably, the nature of the integration and the events leading up to it determine the physiology of the infected cell. A partial insertion can lead to a nonproductive state (no viruses liberated from the infected cell), to an S+L− state (defective viruses liberated from the infected cell), or to a state where recombinant viruses are liberated from the infected cell. The nonproductive state can be associated with transcription and translation of viral genes; whether partial integration can also result in a totally quiescent situation, at least temporarily, is not known but we suspect this is usually the case. It is common for infected cells to liberate defective particles (BASSIN, TUTTLE and FISCHINGER, 1970; BASSIN *et al.*, 1974). It seems that it is also common for retroviruses to acquire cell sequences (SHOYAB, MARKHAM and BALUDA, 1975). Sometimes, these correspond to virogene sequences (HAYWARD and HANAFUSA, 1975) and occasionally they confer transforming potential upon the virus (SCOLNICK *et al.*, 1973; STEHELIN *et al.*, 1976; SCOLNICK, GOLDBERG and WILLIAMS, 1976).

As expected, the infection process can cause changes in the genome of the invading virus (ALTANER and TEMIN, 1970). However, in cells carrying and expressing both complete and defective proviruses, transcripts of the complete proviruses are preferentially packaged into mature virus particles. In this manner, retroviruses are able to preserve some genetic continuity in the face of strong mutational pressure.

The parameters of the virus-infected cell are described above from the viewpoint of the virus. The metabolism of the cell following infection is probably also determined by the nature of the integration event. Whether the cell is transformed certainly depends upon whether *src* genes of sarcoma viruses become integrated.

Since current models for the origination of sarcoma viruses involve the interaction of endogenous viruses or leukemia viruses and cell genes, certain types of infections by these viruses may also transform cells. Whether the infected cell is malignant in the sense of being able to cause tumors in recipient animals may also be a consequence of exactly how the virus integrated, although we are not aware of evidence to support this statement.

It is our bias that the mechanism that viruses use to cause malignancy will not be restricted to them. We believe that malignancy is characterized by a type of genetic change efficiently carried out by RNA tumor viruses, inefficiently caused by mutagenic carcinogens and only indirectly stimulated by nonmutagenic agents. Clouding this relatively simple model are the facts that carcinogens have other effects, that they interact with one another, that susceptibility to these agents is modified by the genetic background and developmental state of the host and that most of our information comes from examining the endproduct of all of these interactions, rather than each alone. By the time the phenotype of the malignant cell is determined, probably by the time its new complement of proteins is synthesized, pleiotropy is evident and similarities in the mechanism two different cells or organs used in arriving at the transformed state are lost. A central question in tumor cell biology today is whether there in fact *is* a common event — integration of a certain sequence, integration in a particular location or transcription of particular genes, etc.—that characterizes malignancy in general.

In this monograph we will make a special effort to place facts relating to the molecular biology of retroviruses and cell transformation in an evolutionary context. It is generally thought that cancers are manifestations of the abortive use of a genetic system and that the origination of retroviruses can be explained similarly. It seems reasonable, considering the pervasiveness of the cancer phenotype and feeling intuitively that the systems of gene mobility and splicing are error-prone, that the genetic system will be newly evolved, not perfected, and easily aborted. In such a context RNA tumor viruses are especially adept at entering and interfering with this putative new genetic system (GILLESPIE, MARSHALL and GALLO, 1972). In the past it has been proposed that this genetic system involves vertebrate development (HUEBNER and TODARO, 1969), reverse transcription and recombination with relocation (TEMIN, 1971), RNA processing (GILLESPIE and GALLO, 1975), growth-regulating proteins, etc. Probably, all of these ideas will have an element of truth. We feel that they will all be aspects of a more-encompassing genetic system. Specifically, we consider the following idea: that the genetic system is the evolution of a dynamic genome from a static (relatively) one. In this context when *int* is a repeated DNA, a newly evolved repeated DNA sequence, the integrated portion of the virus genome would become a moveable gene and might integrate more or less randomly at many points throughout the genome (at related repeated DNA locations). Reintegration at particular sites, possibly at virogene locations, would affect the expression of these sites, leading to altered growth patterns, transformation. In the model, virogenes are not normally moveable genes. When an *int* is brought next to a virogene, the virogene would become mobile, infectious, capable of being transferred to another location in the same cell, to another cell, even to another animal species. Some of the more remarkable aspects of this model, for example, the idea of a gene

becoming infectious and being transmitted to other animal species are well-docu-
mented (BENVENISTE and TODARO, 1974). The more mundane questions—whether
repeated DNA can be *int* sites, if so whether integration at repeated DNA plays
a role in determining an infection leading to transformation, etc.—remain poorly
documented.

Though this article deals chiefly with the information transfer aspect of RNA
tumor viruses, a brief introduction to their structure and molecular components
is useful. The viruses themselves are roughly spherical particles about 1000 Å in

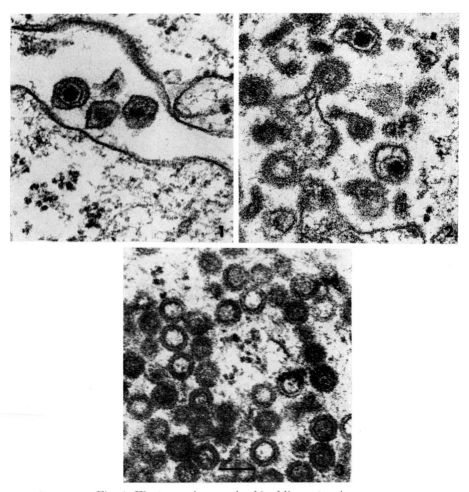

Fig. 1. Electron micrograph of budding retroviruses

diameter composed of a mixture of cellular and viral-specific molecules. The
viruses bud from cell membranes so their outer "coat" is composed primarily of
cell structures (Fig. 1). Within this outer membrane, however, is the major viral
glycoprotein, a molecule of approximately 70,000 daltons, the gp 70. This protein
often helps determine the host range of the virus.

Situated within the 1000 Å particle is a smaller sphere of about 500 Å containing the RNA genome of the virus. The membrane of this core or nucleoid is formed just inside the plasma membrane of the cell during the early stages of virus budding. The major protein of the core membrane is a 30,000 dalton, virus-specific protein of unknown function, the p 30.

Within the core resides the viral RNA and two associated proteins, the reverse transcriptase (60,000 daltons) and the p12, a phosphorylated protein of 12,000 daltons. The reverse transcriptase carries out the function that distinguishes retroviruses from all other viruses, the formation of DNA copies of the RNA genome. The p12 is a protein that specifically recognizes viral RNA but little is known concerning its biological role.

Analogous proteins from different retroviruses vary in molecular weight, but within remarkably narrow limits. Despite considerable amino acid sequence divergence of analogous proteins from, for example, birds and primates there has been considerable conservation of the size of retrovirus structural proteins. To date there is no accepted explanation for the conservation of viral protein size or, conversely, for the lack of sequence conservation of analogous viral proteins.

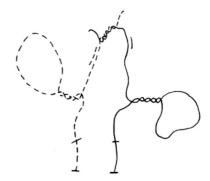

Fig. 2. Structure of Retroviral 30—35 S RNA.

Two large 30—35 S RNA molecules are represented; one by a solid line and the other by a dashed line. The 5′ end is at the top, the 3′ end of the RNA is below. The poly(A) region is bounded by horizontal slashes. Regions of secondary structure are indicated by double helices; only the most stable hydrogen-bonding linkages survive the spreading procedure used for making electron micrographs of the viral RNA. The primers are represented by short lines parallel to the 30—35 S RNA and near the dimer linkage structure at the 5′ end. After KUNG et al. (1976) and BENDER and DAVIDSON (1976)

The retroviral RNA genome is an unusual aggregate of molecules. There are four important RNA molecules in each viral core; two large, probably identical RNA molecules about 10,000 nucleotides long and containing the coding information for the viral proteins and two (probably) identical tRNA molecules less than 100 nucleotides long and serving as "primers" for reverse transcription. The elegant electron microscopy of SILVIA HU and NORMAN DAVIDSON and their coworkers has established the arrangement of the two large molecules as that drawn schematically in Fig. 2.

The large RNAs are joined at their 5' ends in a "dimer linkage" structure. The dimer linkage involves hydrogen bonding of a few tens of nucleotides. In cases where the potential for packaging two different large RNA molecules is generated, heteropackaging is not seen (MAISEL et al., 1978). Thus, there may be nucleotide sequence selectivity in the dimer linkage.

At the 3' end of the large RNA molecules is a long stretch of poly (A). The p12 protein is thought to be bound to the viral RNA at this end while reverse transcriptase binds at the tRNA primer site.

The map of avian virus RNA has been determined and a generalized sarcoma virus map is shown in Fig. 3. Three major loci code for about 7 proteins. The loci have been called *gag*, coding for several low molecular weight proteins, including p 30 and p 12; *pol*, coding for reverse transcriptase; and *env*, coding for envelope glycoprotein, gp70[1]. In each case the locus is translated as an intact precursor

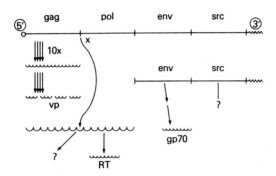

Fig. 3. Map of sarcoma virus RNA.
Genes are abbreviated as follows: *gag* small, nonglycosylated viral proteins such as p 30, p 15, p 12, etc. (vp); *pol* reverse transcriptase (RT); *env* envelope protein (gp 70); *src* sarcoma-specific gene. Translation of 30—35S RNA begins at the 5' end of the RNA molecules. *Gag* is frequently translated into a polypeptide precursor from which the nonglycosylated viral proteins are cleaved and infrequently into a polypeptide precursor containing reverse transcriptase. To translate *env*, it appears that the 30 to 35S RNA must be cleaved and the 3'-proximal portion can then be translated into a polypeptide from which gp 70 is cleaved. *Src* is probably translated from the cleaved RNA but its mechanism, whether it is a read-through product or a specially-processed polypeptide, remains speculative

polypeptide, then mature viral proteins are produced as cleavage products of the precursor (JAMJOOM et al., 1977). Sometimes precursor proteins are packaged into mature virus particles. Occasionally, *gag* and *pol* are transcribed as a single unit (KOPCHICK et al., 1978). *Gag* and *pol* are translated from the large RNA, indistinguishable from virion RNA. *Env*, however, seems to be translated from a message RNA molecule representing the 3' end of the genomic RNA. Some cytoplasmic, viral-specific RNA molecules may be untranslatable (TSUCHIDA and GREEN, 1974).

---

[1] p 12, p 30 and gp 70 proteins of murine retrovirus have analogues of approximately the same size in other retroviruses, e. g. p 15, p 27 and gp 85/gp 37 of avian retroviruses.

Other loci are defined by biological function rather than by coding potential. Avian oncornaviruses possess a "host range" marker near the poly (A) end of the RNA (gp70 also influences the host range of the virus). RNA viruses that transform fibroblast cells in culture carry a *src* sequence; in avian viruses *src* is between *env* and poly (A). *src* reportedly codes for a protein (BEEMON and HUNTER, 1977; PARKS and SCOLNICK, 1977; PURCHIO *et al.*, 1978), but there is little agreement concerning the nature of the transforming protein *in vivo*. The ideas that it is a protein kinase, an unprocessed gp70, an unprocessed *gag* protein, even a modified p 30 have all been proposed within the last year.

Since this review was completed, several investigators used temperature-sensitive mutants of Rous sarcoma virus and Kirsten sarcoma virus to describe these *src* gene products as protein kinases (COLLETT and ERIKSON, 1978; RUB-SAMEN, FRIIS, and BAUER, 1979; SHIH *et al.*, 1979). Additionally, the mouse sarcoma virus *src* protein possesses a guanine nucleotide-binding activity (SCOLNICK, PAPAGEORGE, and SHIH, 1979). In normal chickens, an analogous protein has been discovered (COLLETT *et al.*, 1979). One problem now is whether the *src* protein only causes cell transformation *in vitro* or whether it causes tumors *in vivo*. LAU *et al.* (1979) showed that synthesis of mature *src* protein did not correlate with whether a Rous sarcoma virus-infected (vole) cell line was transformed *in vitro*, but did determine whether the cells would be tumorigenic in nude mice. Conversely, POSTE and FLOOD (1979) claimed that chick embryo fibroblasts infected by a mutant of Rous sarcoma virus induced tumors on the chorioallantoic membrane of chicken eggs even when active *src* protein was presumably not produced. It will be interesting to do these experiments in a biologically homologous system, as did POSTE and FLOOD (1979), carefully measuring *src* protein activity as did COLLETT *et al.* (1979). However, the situation is now complicated by the possibility that there may exist several different *src* genes in any given species (BISTER and DUESBERG, 1979; DUESBERG and VOGT, 1979).

Infection of cells by retroviruses is often abortive, such that only some of the functions specified by the intact viral genome are fixed in the infected cell. It is becoming increasingly popular to consider that the variability in the biology of infected cells results from variability at the integration step. It is no longer thought that the entire virus genome must integrate into the host chromosome. The biology of the infected cell and of the viruses liberated by productively infected cells might be solely determined by which viral sequences become integrated and by their location, *e.g.* by the details of the integration event. It is of some use, therefore, to outline the steps permitting the RNA genome of the virus to interact with the DNA genome of its host.

One of the first events in the infection of a cell by a retrovirus is the synthesis of DNA copies of the viral RNA (Fig. 4). This activity is carried out by reverse transcriptase, an enzyme brought into the cell by the virus. This enzyme is capable of using an RNA template for directing the synthesis of complementary DNA (cDNA), providing that a primer molecule for accepting the first deoxynucleotide is associated with the template.

In the case of the retroviruses, the primer is a tRNA molecule and is located near the 5' end of the RNA template, the 30—35S viral RNA genome (Fig. 5). Reverse

EVENTS IN RNA TUMOR VIRUS INFECTION AND EXPRESSION

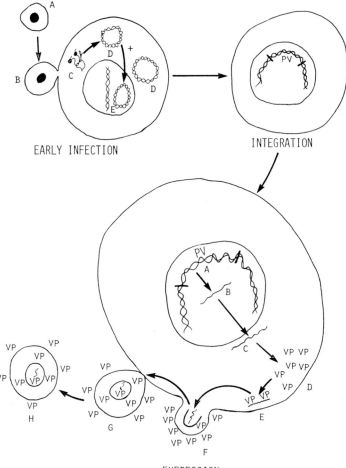

Fig. 4. Life cycle of Retroviruses.

An RNA tumor virus particle *(A)* fuses with a cell *(B)*. The RNA of the particle becomes uncoated *(C)* and is transcribed into a double-stranded free provirus of DNA *(D)* by associated reverse transcriptase (see Fig. 5 for details of provirus synthesis). The double-stranded provirus enters the nucleus *(E)* and becomes integrated into the cell chromosome *(PV)*

Expression of the provirus is schematized in the lower part of the figure. RNA is transcribed from the integrated provirus, processed and transported to the cytoplasm *(A—D)*. It is translated into viral proteins *(VP)*. Viral proteins associate with RNA near the cell membrane *(E)*, form virus "buds" *(F)* and viruses are released into the extracellular fluid *(G, H)*

transcriptase begins synthesizing cDNA in a direction *toward the 5' end!* Reaching the end of the template, the enzyme-cDNA "jumps" across to the 3' end, skipping the poly (A) region, and continues synthesizing cDNA until the initiation site is again reached (TAYLOR and ILLMENSEE, 1975; JUNGHANS *et al.*, 1977). Actually, the enzyme does not jump from one end of the template to the other. A short sequence is repeated at both ends of the RNA template, allowing the structure to circularize once the 5′ end is reached (HASELTINE, MAXAM and GILBERT, 1977; SCHWARTZ, ZAMECNIK and WEITH, 1977). This redundant sequence is called the "a" sequence in Fig. 5 and one copy of it is lost during the formation of cDNA. To make double-stranded, free provirus the cDNA is used as template and a complement of it is synthesized. Apparently, this strand is synthesized a few nucleotides at a time, then oligomers are sewn together (VARMUS et al., 1978). In any event, the double-stranded, free provirus circularizes, becoming a superhelical structure (GUNTAKA et al., 1975; GIANNI, SMOTKIN and WEINBERG, 1975). The superhelical provirus migrates to the nucleus where the integration event takes place.

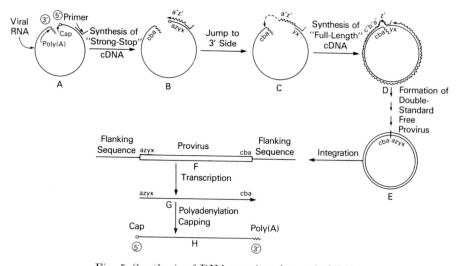

Fig. 5. Synthesis of DNA provirus from viral RNA.
Viral RNA consists of 30—35S RNA, polyadenylated and capped, and associated RNA primer (A). cDNA synthesis begins at primer and moves toward the 5′ end of the 30—35S RNA (B). The cDNA "jumps" to the 3′ end by hybridizing to redundant sequence, "a" (C). The remainder of the 30—35S RNA is copied into complementary DNA (D). A second DNA strand is synthesized, creating a double-stranded provirus, which circularizes (E). The "a" sequence must be duplicated during integration into the chromosome (F) or during RNA synthesis (G) or processing (H). See Gilboa et al. (1979) for recent details

Recent articles by CLAYMAN et al. (1979) and GILBOA et al. (1979) more precisely detail the synthesis of double-stranded proviral DNA by enzymes contained within retrovirus particles.

If aberrant integrations were to take place, they would be likely to result from aborted cDNA synthesis and complete integration of free provirus or from unusual recombination events between the chromosome and a complete or altered provirus. It will be a major task of this monograph to decide whether there is evidence pertaining to this point and to try to correlate types of integration events with the known biological properties of retroviruses, including the ability of some of them to cause cancer in animals.

II. Origin of Retroviruses

It is common to speak of viruses as autonomous genetic elements, dependent on cell machinery for replication, at least in part, but consisting of an RNA or DNA genome unrelated in nucleotide sequence to genes of its host. It does not necessarily follow that virus genomes which are unrelated to cell genes did not originate from them. If the genome of a given virus evolved independently of the cellular genes the degree of virus-cell homology would depend on the rate of virus-cell divergence since the origination event and the elapsed time. RNA tumor viruses offer a unique model for studying the origination of viruses from cell genes, for it appears that many existing RNA tumor viruses originated recently from cellular genes. Indeed, retroviruses can in many instances be derived from uninfected cells in the laboratory.

In 1964, TEMIN proposed that recombination between an infecting RNA tumor virus genome (actually its DNA copy) and cell genes was a required step in the replication of the virus. Testing of the "provirus" theory over the past 14 years has demonstrated its validity. It is now known that recombination between RNA tumor virus genomes and cell genes results in the acquisition of host nucleotide sequences by the virus (SHOYAB, MARKHAM and BALUDA, 1975; HAYWARD and HANAFUSA, 1975). This complicates studies of virus origin, for the presence of host-related sequences might reflect either the genetic origin of the virus or the nature of its recombination events with cells.

In 1969, HUEBNER and TODARO proposed that normal, uninfected animals carry genes capable of giving rise to RNA tumor viruses, genes they called "virogenes". A virogene is defined as a normal cell gene, whereas a provirus is a new gene inserted into the chromosome of a normal cell during the infection process. For some reason it is out of vogue to talk of virogenes but we find the concept useful when considering retroviruses in an evolutionary context. The difference between the provirus and virogene theories is that the provirus theory does not speak to the origin of the virus genome whereas the virogene model specifies that the virus RNA is an exact copy of particular genes in normal cells. It is now known that there are viruses of the type proposed by HUEBNER and TODARO (endogenous viruses) but there is no evidence they cause cancer as the model predicted. Thus, the virogene model is most useful when considering the origin of retroviruses rather than their tumorigenic potential.

In the early 1950s, GROSS accumulated evidence that the capacity of mice to produce certain RNA tumor viruses was a vertically-transmitted characteristic (GROSS, 1951). Later, the presence of viral-related antigens in embryonic chicken and mouse cells was reported (DOUGHERTY and DiSTEFANO, 1966; PAYNE and CHUBB, 1968; WEISS, 1969; GILDEN and OROSZLAN, 1971; see TOOZE, 1977 for references). These observations led HUEBNER and TODARO (1969) and BENT-VELZEN and DAAMS (1969) to speculate that RNA tumor virus genomes originated from cell genes, that these viral genes are under the control of repressors like those postulated by JACOB and MONOD (1961) and that derepression of these genes leads to cancer.

Support for the virogene theory has come from several areas of research. LOWY et al. (1971) and AARONSON, TODARO and SCOLNICK (1971) showed that treatment of normal cells in culture with halogenated pyrimidines, especially iododeoxyuri-

dine, IdU, can cause cells to produce endogenous viruses. The induction is aug-
mented by certain steroids and prevented by inhibitors of nucleic acid metabolism
(WU et al., 1972, 1974). The viruses are probably also induced by infection of cells
by other viruses.

The early induction experiments could have been criticized on the grounds
that the cells used were defined as normal only because they were withdrawn
from apparently normal animals, then cultured. They could have become infected
during the process of culturing. However, more recent studies with primates
indicate that this is not the case. Using the electron microscope, several workers
noticed type C particles budding from cells of fresh placental tissue, especially from
baboon placenta (SCHIDLOVSKY and AHMED, 1973; KALTER et al., 1973). MELNICK
et al. (1973) were able to rescue a new virus from baboon cells by superinfection with
feline sarcoma virus. Subsequently, MELNICK and TODARO collaborated to induce
infectious virus from primary cell strains with IdU (TODARO et al., 1974). The
baboon virus was passed to secondary cells and has been analyzed in several ways.
It is clearly an endogenous virus of baboons (BENVENISTE and TODARO, 1974,
1976; WONG-STAAL, GILLESPIE and GALLO, 1976; DONEHOWER, WONG-STAAL
and GILLESPIE, 1977). Thus, it appears that normal tissues as well as cells
cultured from them are capable of producing retroviral particles.

Molecular hybridization constitutes the main evidence that cell genes can give
rise to retroviruses. Viral RNA or cDNA has been hybridized to DNA from normal
or virus-infected tissue or to DNA from cultured cells. It is assumed that virus-
producing cells carry in their DNA all of the viral sequences found in the viral
RNA genome. RNA or cDNA from some retroviruses hybridizes as well to DNA
from certain uninfected animals as it does to DNA from cells presumably containing a complete provirus. Thus, all the genes of this type of retrovirus can be
detected in DNA of uninfected animals. For example, the RNA of baboon endogenous virus hybridizes as extensively to DNA from normal baboons as it does to
human cells infected by and producing the virus (Fig. 6). By similar tests RD 114
is an endogenous virus of cats (NEIMAN, 1973) and the endogenous guinea pig
virus is a copy of genes in normal guinea pigs (NAYAK, 1974) and several types of
RNA viruses that can be induced from mice are replicas of mouse genes (BENVE-
NISTE et al., 1977).

These experiments have shown that DNA from uninfected tissues of animals
contains sequences complementary to a major fraction, if not all, of the RNA
genomes of certain viruses they liberate. By the molecular hybridization criteria
available, the viral DNA sequences (the virogenes) are the same as sequences
in the viral RNA, not just related to them. It is somewhat vexing that one cannot
routinely hybridize 100% of the RNA of these viruses to DNA from cells. Failure
to do so is usually considered to result from technical limitations of molecular
hybridization with complex DNA genomes; equal hybridization to DNA from
uninfected (virogene) or virus-infected (provirus) cells attests to this interpretation.
The observation that on occasion 100% of a cDNA preparation hybridizes to DNA
of normal tissues is not necessarily comforting because cDNA is seldom, if ever, a
complete representation of its RNA template.

Virogene sequences in animal genomes show unusually rapid evolutionary
change, compared to unique DNA of the same group of animals (GILLESPIE and

GALLO, 1975; BENVENISTE and TODARO, 1976). Therefore, RNA from a given virus hybridizes best to DNA from only one type of animal, the type that created it. In this context, the word "type" means species or genus. Hybridization of the same RNA to DNA from distantly related animals will be less extensive. Thus, RNA or cDNA from the endogenous chicken viruses RAV-0 and reticuloendotheliosis virus hybridize best to DNA from normal chickens and less well to DNA from other birds (NEIMAN, 1973; KANG and TEMIN, 1974; TEREBA, SKOOG and VOGT, 1975; SHOYAB and BALUDA, 1975; but see FRISBY et al., 1979)[1]. These viruses can truly be called endogenous chicken viruses because apparently normal chicken cells liberate them and also because there are genes in chickens identical to the RNA of these viruses (within the experimental limits of the techniques). Similarly, genomes of xenotropic mouse viruses hybridize best to DNA from *Mus musculus*, the species of mouse from which they were obtained, less to DNA from other mouse species or to rats and poorly to DNA from other animals. (CALLAHAN et al., 1974; BENVENISTE and TODARO, 1974). RNA or cDNA from baboon endogenous virus hybridizes best to the baboons, *Papio cynocephalus* and *Papio anubis*, then, in decreasing order of ability to hybridize the RNA: other *Papio* species, mandrill and gelada baboons, mangabeys, macaques, guenons, and primitive old world monkeys and apes (Fig. 6) (BENVENISTE and TODARO, 1976; DONEHOWER, WONG-STAAL and GILLESPIE, 1977). This type of phylogenetic continuum shows not only that virogenes exist in the species of baboon that created the virus *(Papio cynocephalus)*, but also that they are the products of evolution and have been present as long as the primate order has existed. Similar results have been obtained with viruses from other species of animal.

Thus, the RNA of endogenous RNA tumor viruses hybridizes best to DNA of the animal from which the virus was isolated even though the animals studied had no contact with viruses. In some cases the animals were raised in a germ-free environment. In other instances, the virogenes have been found in animals obtained from geographically remote regions. These results seem convincing to us that the genome of this class of RNA tumor viruses is encoded also in the genome of normal progenitor animals and, in combination with the biological experiments, mark these sequences as the virogenes postulated by HUEBNER and TODARO (1969).

So far, all of the RNA tumor viruses of this group, containing RNA that hybridizes to DNA from normal tissue in the manner described above are endogenous viruses and lack tumorigenic activity (but see STEPHENSON, GREENBERGER and AARONSON, 1974) Unfortunately, "endogenous" RNA tumor viruses are usually classified by biological criteria and thus are defined by their phenotype. Classifying the viruses by genotype, e.g. by molecular hybridization, need not group the viruses in the same manner. Some viruses, the AKR mouse leukemia virus, the Kirsten sarcoma virus and some strains of the mouse mammary tumor virus have occasionally been called endogenous, but by molecular hybridization they cannot be placed in this category because they carry sequences not found in normal animals (GILLESPIE and GALLO, 1975; CHATTOPADHYAY et al., 1974). An endogenous virus will be defined here as one whose RNA genome is encoded

---

[1] The age of the RAV-0 virogene is uncertain because we cannot decide whether it was recently introduced by infection of chickens (FRISBY et al., 1979) or whether it is an ancestral, rapidly-evolving chicken gene (see pp. 25, 71).

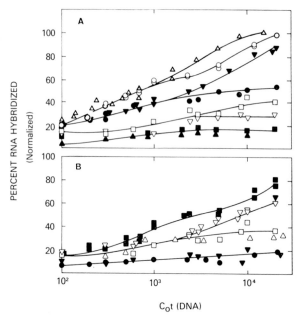

Fig. 6 A and B. Hybridization of RNA from baboon endogenous virus to DNA from primates.

Approximately 1,500 cpm of ^{125}I-labeled RNA (0.015 ng) was mixed with 50 µg of DNA in 10 µl of 0.4 M phosphate buffer. Samples prepared in replicate were heat-denatured, hybridized at 60° for particular lengths of time, and assayed for formation of structures resistant to ribonuclease A. Each C_0t unit corresponds to 0.01 hr. Animals or cells used to prepare DNA were as follows: *A* △ Papio cynocephalus; ○ Papio anubis; ▼ Papio papio; • Mandrillus sphinx; □ Pan troglodytes (chimp); ▽ Hylobates lar (gibbon); ■ Rattus rattus (rat); ▲ Mus musculus (mouse). *B* ■ NC37 (BaEV) cultured human lymphoid cells producing BaEV virus; ▽ Cercocebus torquatus (mangabey); □ Macaca cyclopis (macaque); △ Felis catus (cat); ▼ Presbytis cristatus (langur); • NC37 uninfected cultured human cells. Results are normalized to the maximum percent of the RNA hybridized to DNA from Papio cynocephalus (50%). From DONEHOWER, WONG-STAAL, and GILLESPIE (1977)

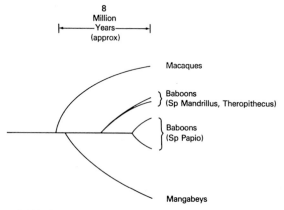

Fig. 6C. Phylogeny of the baboon-macaque-mangabey group.

in toto and exactly in DNA of every tissue of every member of a phylogenetically-defined group of animals, *e.g.* a species. For this reason the term class 1 virus was coined to describe an RNA tumor virus containing a genome that hybridizes equally from uninfected or virus-infected cells.

Since TEMIN's original provirus paper in 1964 it has been known that some tumor virus genomes contain genes not present as such in the chromosomes of the animals from which they were obtained, a result inconsistent with the simplest form of the virogene theory. In 1971, TEMIN proposed an idea reconciling the presence of novel sequences in viruses with their apparent cellular origin. He said that even though some RNA tumor viruses appear to have originated directly from genes of normal cells, these viruses are not tumorigenic and that genetic change is required to produce a tumorigenic virus. This "protovirus" theory put forth a specific mechanism for promoting this genetic change, a mechanism involving repeated cycles of copying virogene RNA into free provirus which then recombined with the cell chromosome. In practice, it has been found that the tumorigenic viruses all carry sequences not found in the genome of the natural host animal. This is seen experimentally as more hybridization of the viral RNA to DNA from virus-infected, virus-producing cells than to DNA from normal tissue. These viruses are called "exogenous" or class 2 viruses. Class 2 viruses include chicken sarcoma and leukemia viruses, mouse leukemia and sarcoma viruses, cat leukemia and sarcoma viruses, bovine leukemia virus, and primate sarcoma and leukemia viruses (SHOYAB, BALUDA and EVANS, 1974; SHOYAB, EVANS and BALUDA, 1974; SHOYAB and BALUDA, 1975; GILLESPIE, *et al.*, 1973; GILLESPIE and GALLO, 1975; LONI and GREEN, 1975; SCOLNICK, *et al.*, 1974). In each of these cases when viral RNA is used as a probe and especially when hybridization conditions are adjusted to discourage the formation of poorly matched complexes only 5—20% of the viral RNA hybridizes to DNA from tissues of normal members of the natural host species (GILLESPIE, *et al.*, 1973).

It should be emphasized that hybridization results obtained using cDNA synthesized by some of the viruses are contrary to those obtained using RNA. The cDNA synthesized by the viruses mentioned above hybridizes nearly as much to DNA from normal animals as it does to DNA from cells infected by and producing the respective viruses, though the hybrids are not as perfectly matched (see GILLESPIE, SAXINGER and GALLO, 1975, for review). This probably means that the cDNA used for those experiments was copied from the 5—20% of the RNA that is most closely host-related, the ends of the RNA molecule (Fig. 5).

An important example of this discrepancy is encountered with feline leukemia virus. This virus became widespread in the cat population some 3—5 million years ago (BENVENISTE, SHERR and TODARO, 1975) and causes leukemia in that species (see ESSEX, 1975, for review). BENVENISTE and TODARO claimed that the exogenous viral sequences entered the germline of cats and that the virus is now endogenous since cDNA copies of the virus RNA hybridized as well to DNA from normal cats as it did to DNA from cells infected by and producing the virus. Results with virus RNA deny this conclusion (KOSHY *et al.*, 1979). Only 20% of the RNA hybridizes to DNA from normal cats while over 50% hybridizes to DNA from cells producing the virus. The virus is clearly a class 2 virus and cannot therefore be considered endogenous. We believe that the virus was once endogenous to cats but that it has since evolved away, becoming a class 2 virus. The origin of feline leukemia virus is treated in detail in the Section V (Transmission of Retroviruses Among Animals) and the genetic relationship with its host parallels that between Rauscher leukemia virus and its (mouse) host, described below.

The observation that the bulk of the sequences in the RNA of class 2 viruses does not hybridize stably with DNA of normal animals shows quite clearly that the genome of these viruses is not coded *per se* in genes of cells. However, under

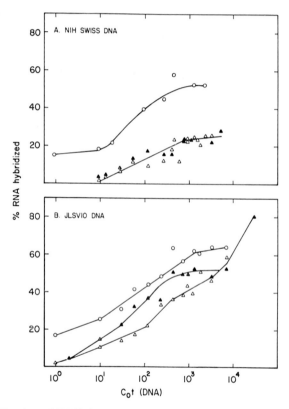

Fig. 7. Hybridization of RNA from Rauscher Leukemia Virus to DNA from mice. Hybrids containing 30 μg of RNA and 50 μg of DNA were prepared in replicate and incubated in 0.4 M phosphate buffer at 60° or 70°. Aliquots were taken at various times and RNA · DNA complexes were detected by binding to nitrocellulose (detects all complexes) or by resistance to ribonuclease A (detects the most perfectly matched complexes). ○ nitrocellulose assay on hybrids formed at 60°; △ ribonuclease assay on hybrids formed at 60°; ▲ ribonuclease assay on hybrids formed at 70°. *A:* normal mouse DNA; *B:* DNA from mice producing Rauscher leukemia virus. From GILLESPIE, GILLESPIE, and WONG-STAAL (1975)

conditions of molecular hybridization that favor the formation of hybrids between related nucleic acids, in addition to identical ones, it can be shown that the *bulk* of the sequences in RNA of some class 2 viruses is related, though distantly, to DNA sequences in normal animals.

The simplest case of this type is Rauscher mouse leukemia virus (GILLESPIE, GILLESPIE and WONG-STAAL, 1975). Only 20% of the RNA of this virus forms RNA-DNA complexes with DNA from normal mice that withstand exposure to ribonuclease A in 0.3 M NaCl at 37° C. These conditions of hybrid detection will detect the interactions of nucleic acids that are more than 85% related to one another, judging from the thermal stabilities of the hybrid structures. Are there also DNA sequences in the mouse related to the genome of Rauscher leukemia virus, but less so than the homology detected by the ribonuclease assay? To test this one would have to score *all* hybrid structures, not just those resistant to nuclease. This can be done by trapping the DNA molecules and associated RNA on nitrocellulose filters. When this assay is performed on hybrids formed between RNA from Rauscher leukemia virus and DNA from normal mice (Fig. 7), it is found that the majority of the RNA is converted to hybrid with DNA, as much in fact as when DNA from virus-producing cells is used (the JLSV 10 cells of Fig. 7). The hybridization is a specific one, since it is not obtained when DNA from humans or rats is substituted for the mouse DNA, but the relation between the viral RNA and the cell DNA is fairly distant with some of the hybrids being 30% or so mismatched, from thermal denaturation profiles. Thus, the relationship between the RNA of Rauscher leukemia virus and the DNA of normal mice is such that some of the regions of the RNA are mirrored by closely related mouse DNA sequences while other parts of the RNA are not as closely matched by sequences in the mouse genome. This situation would follow normally if the Rauscher virus originated from mouse cell genes but evolved away, some regions of the RNA accumulating changes more rapidly than others.

As far as is known, Rauscher leukemia virus has been grown only in mice and has not replicated in other animals, except experimentally. It is not likely that the Rauscher virus has sequences derived from any animal but mice; indeed the Rauscher virus sequences only distantly related to mouse genes are more related to genes in mice than to genes in other animals as if even they were originally derived from mouse genes (see also Section VI on Relatedness Among Retroviruses). How then could the sequences in Rauscher virus become less and less like the original progenitor gene? The only reasonable answer seems to be TEMIN's (1971); free proviruses in the cytoplasm of cells can recombine with chromosomes at new sites, sites slightly different from the gene that served as the original template for the provirus. In Temin's formulation the original provirus, an *exact* copy of a virogene, is called a protovirus and it is not until enough genetic change occurs that the virus escapes the controlling host regulatory mechanisms that it is properly called a provirus.

Rauscher virus, then, is an example of an RNA tumor virus that has apparently stayed within a species and evolved away from its original progenitor genes. Feline leukemia virus is an example of an RNA tumor virus that was transmitted from one animal species to another and is evolving in its new host. Both Rauscher and feline viruses are tumorigenic while their progenitor endogenous virus homologues are not. These viruses and several others like them provide indirect support for the protovirus model, both in terms of explaining how a host gene can escape normal regulation and also supporting the proposal that the property of oncogenesis requires genetic change of the virus after its origination.

As we have defined them, class 2 viruses carry a majority of sequences not found in normal progenitor animals while class 1 viruses carry genomes that are exact copies of normal genes. The criteria are established by molecular hybridization, though the classifications are roughly equivalent to exogenous and endogenous viruses, a biological definition. Some viruses are intermediate between class 1 and class 2 in hybridization properties in the sense that only a minority of their sequences are different from normal cell genes. The AKR mouse leukemia virus, for example, is called endogenous by many because it can be routinely isolated from the AKR strain of mouse, a strain with a high incidence of leukemia (GROSS, 1951; ROWE, 1972). A complete copy of the virus genome can be found in this mouse strain and in other related, high incidence of leukemia strains. However, the virus is not an endogenous mouse virus because most mice do not contain the *entire* virus genome in their DNA (CHATTOPADHYAY et al., 1974). The AKR mouse could be a genetic variant of *Mus musculus* carrying a mutant virogene. In this case the mouse strain is probably a laboratory artifact; high-incidence leukemic strains with shortened life spans would hardly be selected under natural conditions. Alternatively, the AKR virus has inserted its novel sequences into the germline of the AK mouse. In this case the virus arose by infection rather than by "natural" selection, as in the first case, and will only become endogenous if the AKR mice and/or its relatives with a high incidence of leukemia are selected for and establish a new species. In both cases, there existed a single progenitor of all mouse strains with a high incidence of leukemia. We have been unable to discern whether this is the case.

We could consider the AKR sequences to be virogene, rather than proviral sequences of the AKR mouse. This definition would seem to be at odds with our tenet that endogenous viruses cannot cause cancer while foreign viruses can. Cancer-causing genes should not be evolutionarily preserved in the germline. However, if virogenes are used during development they are probably selected in a very specific context; in the context of critical evolutionary functions. In a setting lacking outbreeding and selection for fitness, mutations that would normally be disastrous can be perpetuated. The AKR mouse is a widely used strain and is propagated for its abnormality.

Similarly, mouse mammary tumor virus is considered to be an endogenous virus by several workers. There are many strains and not all have been adequately studied. The DW strain carries RNA that hybridizes as completely to DNA from normal mice as to DNA from mice producing the virus (GILLESPIE et al., 1973; GILLESPIE and GALLO, 1975) but the hybrid with DNA from normal mice is not a perfect one, it has a low thermal stability. Thus, the sequences in the virus are subtly different from the sequences in DNA of normal mice. Originally, we placed this virus in the class 1 group but probably there should be a new classification for viruses like AKR and MMTV, one between class 1 and class 2.

The distinction is more than semantic. Class 2 viruses arise from class 1 viruses (GILLESPIE et al., 1975) and in this context the intermediate class may represent viruses in transit from one class to another. The transition is correlated with the acquisition of oncogenic potential and is accomplished by genetic change. A fundamental question, largely unanswered, is how much and what kinds of change are required for these viruses to become cancer-causing? TEMIN (1974) has pointed out that with the avian sarcoma viruses the degree of oncogenesis is related to the extent of genetic divergence from the original host genes, perhaps indicating that random changes promote oncogenicity. Conversely, the *src* gene of avian sarcoma viruses is a hereditable genetic entity, suggesting that directed changes can also be effective in establishing oncogenic potential.

Considering the logic of the provirus, virogene and protovirus theories there are two mechanisms one might consider for virus origin (Fig. 8). The exact meanings of the words "endogenous" and "exogenous" in the context of this

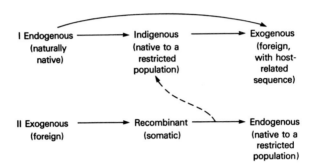

Fig. 8. Models for origin of retroviruses.
Retroviruses can originate through two separate routes:
I. Naturally-selected genes are expressed and direct the formation of endogenous retroviruses. To form indigenous viruses, the endogenous viruses "reintegrate" into the (germline of the) same host species but the virus is genetically modified during the process; simultaneously, the individual carrying the modified virus becomes sequestered. An example is the AKR virus from AK and related high incidence leukemia mice. Reintegration and modification recur in inbred or outbred populations until the exogenous virus is significantly different from the endogenous virus
II. A virus liberated from one species is transferred to a new host species. It remains somatic, not integrating into the germline. Gradually, this exogenous virus becomes adapted to the new host so that infection becomes widespread. This may occur by forming a recombinant virus with regions of homology with the host. Ultimately, the virus titer in the population becomes so high that the probability of a germline infection is significant. If the infected individual is selected for and creates a new species the virus becomes endogenous to that species. Otherwise the virus would be indigenous to a limited group within the species

monograph are as follows: Endogenous viruses arise directly from virogenes and lack tumorigenic potential; foreign viruses (exogenous) contain genomes that are substantially different from any genes in animals and they are in general frankly oncogenic; and indigenous viruses are between—they contain some "novel" sequences and they are mildly oncogenic. There is no conceptual reason that indigenous viruses have to be oncogenic; it is interesting that in fact most of them are oncogenic.

In Fig. 8 normal cell genes, virogenes, are considered to be progenitors of endogenous viruses which, in turn, can become indigenous and then exogenous and frankly tumorigenic (scheme 1). The transitions from one class to another represents accumulated genetic change. The distinction among classes is artificial; RNA tumor viruses represent a spectrum that is continuous from unaltered virogene products to completely foreign genetic elements. The second scheme represents a different interpretation of the same theories. A virus from an unknown source would enter an animal and by recombination acquire hostlike sequences. Having become adapted to the new species it becomes widespread and by further recombination becomes more hostlike. Finally, its sequences are introduced into the germline of the animal, thereby creating a new virogene in the species. Though this manner of introducing a vertically-transmitted virogene into a new species may seem unlikely, even fantastic, BENVENISTE and TODARO (1974—1978) have gathered an enormous amount of evidence that can be interpreted in no other manner (see Section V on Transmission of Retroviruses Among Animals). The introduction of RD 114 virogenes into cats from baboons is the best-documented example of interspecies transfer of virogenes. The model proposed above for the origin of the AKR virus also fits this category.

The two schemes pictured in Fig. 8 are not mutually exclusive. Virogenes created by scheme 2 can serve as progenitors of new viruses and any of the viruses created by scheme 1 might cross species barriers to create new virogenes.

Many of the concepts presented in this chapter ran counter to accepted dogma when they were originally described, but most are accepted now as reasonable working hypotheses from which to proceed. They raise questions and provide approaches to problems of a more fundamental nature, some of which are presented or outlined below.

Are virogenes used during normal development? The presence of viral-related antigens in normal mouse and chicken embryos (DOUGHERTY and DiSTEFANO, 1966; PAYNE and CHUBB, 1968; WEISS, 1969) was one fact that HUEBNER and TODARO (1969) used to support their virogene model. Whether these antigens are products of virogenes is still not known, though evidence has been accumulating that the gp70 coded by virogenes is expressed during differentiation and that different gp70 genes seem to be expressed in different situations (ELDER et al., 1977).

Can "switching-on" of genes cause transformation? This has been an extremely difficult question to resolve experimentally. The central questions are whether oncogenic sequences exist at all, whether they are carried by normal cells and whether they can cause transformation when expressed. If the transformation of fibroblast cells in tissue culture is an acceptable model for oncogenesis, then potentially oncogenic sequences must exist. Rous sarcoma virus of chickens carries a genomic sequence some 1500 nucleotides long that is required for virus transforming potential and is not found in related but nontransforming viruses (STEHELIN et al., 1976). The DNA of normal chickens does not contain this *src* sequence, but it does have a closely related sequence, called *protosrc*. A similar situation holds for transforming viruses of mice and rats (SCOLNICK et al., 1974). Therefore, one could argue that the *src* viral sequences arise by acquisition and modification of cell oncogenes or protooncogenes. In this scenario the cell oncogenes must be genetically modified before they can cause cancer.

We are not necessarily driven to this conclusion, however. Rous sarcoma virus is a highly purified laboratory isolate and cannot be considered representative of chicken sarcoma viruses any more than λdg can be considered representative of λ bacteriophage. *src* can promote transformation but it may not be an exclusive or even common route to chicken sarcomas in the same way that λ virus can incorporate the *gal* genes of *E. coli* but only rarely does. In fact, other chicken sarcoma virus isolates have different *src* sequences. Perhaps *src* acts as a trigger for the induction of normal cell oncogenes or virogenes that in turn cause the oncogenic response.

Transformation by *src* in the Teminesque sense, *i.e.* by integration of a genetically foreign sequence, would not be expected to be reversible. BEATRICE MINTZ has provided strong evidence that mouse teratocarcinoma is a reversible phenotype. She injected teratoma cells into a blastocyst from a normal mouse and allowed the blastocyst to develop within a normal mouse. The teratoma cells became part of the normal tissues of the newborn mouse (MINTZ and ILLMENSEE, 1975; ILLMENSEE and MINTZ, 1976; DEWEY *et al.*, 1977). MINTZ and her coworkers concluded that the participation of the teratocarcinoma cell in a normal developmental program signified that the teratoma phenotype was reversible hence was not a stable character.

The problem with this conclusion is that it runs counter to conclusions from other systems in which the tumor characteristic is a hereditable trait. One solution is to consider the teratocarcinoma a special, atypical case. Another is to specify that genetic change is required for cancer but that it is suppressible under certain conditions. Indeed, in cell fusions the malignant phenotype is usually recessive. Perhaps teratocarcinomas are suppressible by an early developmental activity. This would not be an unprecedented conclusion. Leukemia, a malignancy associated with chromosomal abnormalities, consists of a phenotype (inability of blood cells to differentiate) that can sometimes be passed to daughter cells in culture so it is in that sense hereditable. Nevertheless, the block in differentiation can be reversed by specific growth factors. Thus, it is still reasonable to consider cancer to be genetic change, with the qualification that it be suppressible under certain conditions. With this bias, we turn to an examination of the organization of the genes that are most likely to be involved in this developmental-malignant process, the virogenes.

III. Organization of Endogenous Retrovirus Genes

In mammals virogenes are multiple copy elements. Estimates of the number of copies lie between five and several hundred. This wide range results from variations in the copy number of different virogenes and from technical considerations. The conventional means for determining the number of copies of a particular gene in mammalian DNA is to follow the kinetics of hybridization of labeled RNA or cDNA copies of the gene to a vast excess of unlabeled cell DNA. Under this condition the rate of hybridization of the labeled probe, or more properly the half-life of the unreacted probe, is dependent on the concentration of complementary sequences in the cell DNA (BRITTEN and KOHNE, 1968). The condition under which half of the viral probe hybridizes, the $C_0t\frac{1}{2}$, is related to the number of copies of virogenes per haploid genome. Actually, assigning a copy number depends on knowing: (1) the degree of divergence between the viral probe and the virogenes, (2) that the viral probe does not react with nonviral sequences in the cell DNA, and (3) the kinetics of hybridization of a single-copy standard.

RAV-0, an endogenous virogene of chickens, appears to be single copy, judging from the rate of hybridization of labeled RNA from RAV-0 virus to chicken cell DNA (NEIMAN, 1973). Virogenes in several mammalian species are multiple copy (GILLESPIE, et al., 1973; BENVENISTE et al., 1974). Fig. 6 shows the type of hybridization kinetic data used to estimate virogene copy number (from DONE-HOWER, WONG-STAAL and GILLESPIE, 1977). The $C_0t\frac{1}{2}$ of hybridization of RNA from the baboon endogenous virus to DNA of any of several species of baboons is around 200, whereas the hybridization of RNA to single copy DNA is 2000 (not shown in the figure). There are about 10 copies of virogenes related to the genome of the baboon endogenous virus. Similarly, there are about 10 copies of the virogenes in mice related to the common type-C laboratory viruses and about 10 copies of the virogenes in cats related to the RD114 virus (BENVENISTE and TODARO, 1974). Some experiments with RNA probes yielded somewhat higher copy numbers in these two cases; closer to 50 copies (GILLESPIE et al., 1973; GILLESPIE, GILLESPIE and WONG-STAAL, 1975) but estimates this high are in the minority.

Ideally, one would want stoichiometric evidence for virogene copy number to supplement the kinetic experiments, since interpretation of hybridization kinetics depends on several simplifying assumptions that may not be valid. Three types of stoichiometric molecular hybridization experiments have been used to estimate copy number; DNA saturation, DNA titration and RNA competition. These methods have been treated from a technical point of view elsewhere (GILLESPIE, GILLESPIE and WONG-STAAL, 1975). A DNA saturation experiment is performed by increasing the input of viral RNA or cDNA probe while holding the input of cell DNA constant, then measuring the quantity of viral probe the cell DNA will accept, maximally. The assay is difficult to interpret if the viral probes are contaminated with even traces of nonviral RNA. Accordingly, this assay has not been successfully used.

A DNA titration experiment is carried out by holding the input of viral probe fixed and varying the input of cell DNA, keeping the C_0t value constant. As more DNA is added more RNA is hybridized, until all of the sequences in the RNA complementary to the virogenes in the cell DNA have been taken up into hybrid structures. The approach to completion of the reaction is related to the concentration of virogenes, i.e. to virogene copy number. Usually, a double reciprocal representation of the results is graphed because it yields a straight line when percent of the RNA hybridized is plotted against DNA input and the slope is a linear function of virogene copy number. The results obtained with RNA from baboon endogenous virus are consistent with 10 virogene copies per haploid genome of baboons (DONEHOWER, WONG-STAAL and GILLESPIE, 1977).

Both the hybridization kinetics and DNA titration experiments measure DNA sequence concentration. If one is dealing with a situation where the entire viral probe can be described as complementary to a set of DNA sequences with every member of the set having the same reiteration frequency, then the results take a fairly simple form. A C_0t representation of hybridization kinetics in the simplest case is a sigmoidal curve whose slope at the inflection point spans two logs of

C_0t in its rise from 0 to maximal hybridization (BRITTEN and KOHNE, 1968). Hybridization of RNA from RD114 virus to DNA from normal cats takes this form as shown in Fig. 9A. This, and the rather sharp thermal denaturation profile of the resultant hybrids was interpreted to mean that the multiple copies of the RD114 virogenes in cats were closely related to one another (GILLESPIE and GALLO, 1975). At about the same time evidence was obtained suggesting that RD114 virogenes arose in cats by interspecies transfer of a virus some 3—5 million years ago, rather recently in evolutionary terms (BENVENISTE and TODARO, 1974 and see Section V on Transmission of Retroviruses Among Animals).

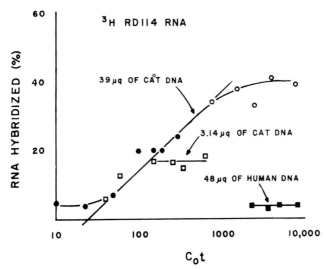

Fig. 9A. Hybridization of RNA from RD114 virus to DNA from a domestic cat. Approximately 300 cpm of ^{33}H-labeled RNA (15 ng) was mixed with the specified amount of DNA in 10 µl of 0.4M phosphate buffer. Samples prepared in replicate were heat-denatured, hybridized at 60° for particular lengths of time and assayed for the formation of hybrids resistant to ribonuclease. From GILLESPIE et al. (1973)

Not all virogenes exhibit this molecular hybridization behavior. The hybridization of RNA from the baboon endogenous virus to DNA from baboons shows complicated kinetics (Fig. 6). Treatment of the results with the Wetmur-Davidson formulation shows that the hybridization kinetics reflects a mixture of reactions with different rates, indicating the participation of cell DNA sequences of different reiteration frequencies (DONEHOWER, WONG-STAAL and GILLESPIE, 1977). Thus, different parts of the baboon virogene appear to have different reiteration frequency. Similar results have been obtained with virogenes in mice (GILLESPIE, GILLESPIE and WONG-STAAL, 1975) and guinea pigs (NAYAK and DAVIS, 1976), both studies using RNA probes.

The notion that baboon virogenes are loci with apparently mixed copy number is supported by DNA titration measurements. As mentioned earlier, a double reciprocal plot of a DNA titration will be a straight line if the set of DNA sequences complementary to the viral probe consists of members with the same repetition frequency. Double reciprocal plots of DNA titration assays using RNA from baboon endogenous virus are two or more lines, depending on the species of baboon used as a source of DNA (DONEHOWER, WONG-STAAL and GILLESPIE, 1977).

Fig. 9B describes a simple interpretation of these results. The basic idea is that early in evolution the multiple copies of virogenes are identical to one another

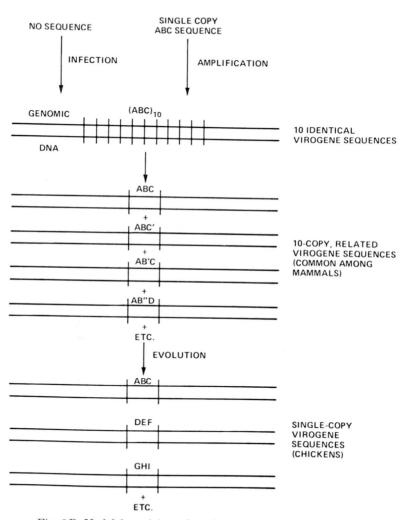

Fig. 9B. Model for origin and evolution of virogene sequences.
It is assumed that virogenes arise by infection or natural selection but in either case they first exist as single copy sequences. They are amplified to 10 or so copies per haploid genome and at the time of amplification all members are effectively identical and probably in tandem. With time the members accumulate mutations giving rise to a family of related, nonidentical genes. Eventually, enough changes are accumulated so that each member appears distinct by molecular hybridization

but with passage of time each copy accumulates genetic changes. Furthermore, since different portions of the virogenes code for different functions they are constrained differently. Some portions will accumulate mutations readily, other regions will be conserved. In the laboratory the RNA probe from a cloned virus represents a copy of only one of the virogenes. The apparent number of copies of the conserved regions of virogenes is higher than the apparent number of the regions that accumulate changes rapidly. Experiments with cDNA tend to mask the differences because cDNA is transcribed preferentially from the conserved portion (GILLESPIE, SAXINGER and GALLO, 1975); nevertheless, the conserved region of baboon virogenes evolved at about 5 times the rate of most single-copy baboon DNA (BENVENISTE and TODARO, 1976). In general, virogenes seem to evolve more rapidly than most cellular genes.

Eventually, members of a multiple-copy virogene family may diverge to the extent that they would appear single-copy by molecular hybridization, as does RAV-0 in the chicken genome (see footnote, p. 13).

The analysis of multiple-copy, divergent genes poses interesting problems not encountered in studying the organization of single-copy genes or multiple-copy, conserved genes like the ribosomal or histone genes. Specific features that characterize one virogene member, features like base sequence or position of cleavage sites for restriction endonucleases, may not be typical of virogenes as a collective family. Conversely, it is not clear how to study features of a particular virogene in the presence of related, nonidentical members or how to purify a virogene of special interest. For example, the gene in AKR mice that codes for the AKR virus is now being cloned in *E. coli*. The AKR mouse also carries multiple copies of virogenes coding for endogenous type-C viruses, non-leukemogenic viruses closely related to the AKR virus. Indeed it is not clear that the "AKR virogene" can be reliably distinguished from the endogenous virogenes.

At an even more subtle level it will probably become important to clone individual virogenes corresponding to N-tropic, B-tropic viruses (see Section VI on Relatedness Among Retroviruses). There is simply not enough knowledge about these virogenes to do this efficiently.

An alternative to examining specific features of virogenes for the purpose of understanding their organization in chromosomes is to explore general features that are independent of the base sequence of the gene. Certain of these features can be compared with properties of the viral RNA genome. Virion RNA is about 10 kilobases in length and has a guanine + cytosine content of about 50% (Table 1 and references cited therein). These features apparently characterize retrovirus genomes as a class; they do not vary substantially among viruses from widely divergent species. The base compositional feature, especially, is a useful one because the bulk of vertebrate cell DNA has a guanine + cytosine content of 40%, so if the base composition of virogenes is like that of viral RNA, virogenes can be separated from the bulk of the remainder of the cell genome.

The high guanine + cytosine content of retroviral genomes is particularly useful for asking whether virogenes are continuously codogenic or whether they are interrupted by noncoding "intervening" sequences. Several cell genes are known to contain internal intervening sequences. Hemoglobin genes (JEFFREYS and FLAVELL, 1977; TILGHMAN *et al.*, 1978), immunoglobin genes (BRACK and TONE-

Table 1. Base composition of RNA from retroviruses[a]

RNA	Host	C	A	G	U	C+A	C+G	C+U	Reference
Avian myeloblastosis virus	chicken	23.0	25.3	28.7	23.0	48.3	51.7	46.0	ROBINSON and BALUDA, 1965
Rous associated virus	chicken	24.2	25.1	28.3	22.4	49.3	52.5	46.6	ROBINSON, PITKANEN and RUBIN, 1965
Rous sarcoma virus	chicken	24.2	24.8	29.2	21.7	49.0	53.4	45.9	ROBINSON, 1967
Rauscher leukemia virus	mouse	26.7	25.5	25.1	22.7	52.2	51.8	49.4	DUESBERG and ROBINSON, 1966
Mouse mammary virus	mouse	21.6	19.3	30.2	28.9	40.9	51.8	50.5	LYONS and MOORE, 1965
Simian sarcoma virus	primate	26.5	26.0	26.5	20.8	52.6	53.1	47.4	JENSIK et al., 1973
Feline leukemia virus	cat	26.7	28.4	24.1	20.8	55.1	50.7	47.5	JENSIK et al., 1973
RD114	cat	22.6	29.9	26.2	21.3	52.5	48.8	43.9	ROY-BURMAN and KAPLAN, 1972
Visna virus	sheep	28.0	23.0	23.0	26.0	51.0	51.0	54.0	LIN and THORMAR, 1971
MEAN		24.84	25.27	26.81	23.07	50.10	51.65	47.91	
VARIANCE[b]		4.38	8.13	5.38	6.46	14.67	1.70	7.99	

a When several estimates for base composition were available, the determination with the lowest standard error was used.

b Variance = s^2

GAWA, 1977) and ovalbumin genes (BREATHNACH, MANDEL and CHAMBON, 1977) in adult cells are interrupted by noncoding sequences several times the size of the coding regions. If virogenes were interrupted by large intervening sequences with guanine + cytosine content of 40%, the average base composition of virogenes would differ markedly from virion RNA. On the other hand, if virogenes were interrupted by large intervening sequences of 50% guanine + cytosine content, the length of the high G + C virogene unit would be considerable larger than 10 kilobase pairs.

The guanine + cytosine content of DNA fragments can be determined by centrifugation to equilibrium in density gradients of NaI. Such a determination, using cat DNA, is pictured in Fig. 10. In this experiment, DNA fragments about

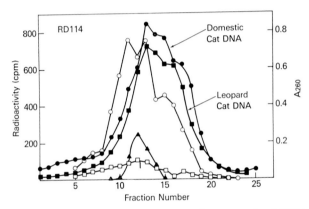

Fig. 10. Buoyant density of RD 114 virogenes of cat DNA.
Centrifugation to Equilibrium. A solution of DNA in NaI was made by mixing 6.0 ml of saturated NaI, 0.5 ml of 0.1 M Tris and 0.1 M EDTA (pH 8.0), 0.1 ml of 1 mg/ml ethidium bromide and 3.5 ml of DNA in 0.015 M NaCl. DNA fragments were reduced in size to an average of 5.5 kilobase pairs by passing the solution 5 times through a $1^1/_2$ inch 27 g needle at 35—40 ml/min. ^{14}C-labeled *E. coli* DNA was added as a density marker, then the final refractive index was adjusted to 1.4340. Ten ml of solution and 2 ml of mineral oil was centrifuged for 90 hr at 25° and 38,000 rpm in a type 40 rotor. Approximately 30 fractions were collected from the bottom of the tube. *RNA-DNA Hybridization.* The density gradient fractions were incubated at 100° for 15 min, cooled in an ice slurry, diluted with 5 ml of ice cold 6 × SSC (SSC = 0.15 M NaCl and 0.015 M Na citrate) and passed through 13 mm filters of nitrocellulose (Schleicher and Schuell). The DNA-containing filters were washed with 15 ml of 6 × SSC and 10 mm circles were cut from the center of each filter. They were then baked for 4 hours at 80° under vacuum. Each filter was incubated at 37° for 60 hours with 5000 cpm of ^{125}I-labeled RD 114 RNA in hybridization solution (50% formamide, 3 × SSC, 0.02 M Tris, pH 7.4, and 0.5% SDS, final concentrations) After hybridization, filters were batch-washed exhaustively with 3 × SSC, then incubated for 2 hours at 37° with 25 µg/ml of ribonuclease A in 3 × SSC and rewashed with 3 × SSC. ^{125}I radioactivity was measured in a gamma counter. DNA was released from the filters by incubation for 40 min at 70° in 1 M HCl. DNA concentration was measured spectrophotometrically; ^{14}C was counted in a scintillation counter in Aquasol

The results of two NaI gradients are superimposed; one shows hybridization to domestic cat DNA, the other to leopard cat DNA. Leopard cats lack the RD 114 virogenes. RD 114 RNA hybridized (cpm) per fraction to domestic (o—o) or leopard (□—□) cat DNA; domestic (•—•) or leopard (■—■) cat DNA released from each filter (A$_{260}$); ^{14}C-labeled *E. coli* DNA marker (cpm, ▲—▲)

half the length of the viral genome were used. Virogene-containing DNA fragments were located in the NaI gradient by hybridization to RNA from the endogenous cat virus, RD114. Fig. 10 shows that the RD114 virogene sequences were recovered from the gradient at a modal position corresponding to 50% guanine + cytosine content, the position where *E. coli* DNA (G + C = 50%) bands. The bulk of the cat DNA was found further up the gradient at the position expected for DNA with a guanine + cytosine content of 40%. The cat DNA was heterogeneous with respect to guanine + cytosine content of DNA fragments in the 5 kilobase pair range, judging from the broad buoyant density profile. *E. coli* DNA banded more sharply because its guanine + cytosine content is conserved along the length of its chromosome (ROLFE and MESELSON, 1959; SUEOKA, MARMUR and DOTY, 1959).

Intervening sequences of 40% guanine + cytosine content would have lowered the *modal* buoyant density of the RD114 virogenes below that actually observed. Probably there are no low G + C intervening sequences 1 kilobase pair or longer near the middle of the RD114 virogenes. Flanking low G + C regions would lower the buoyant density of terminal virogene fragments, an effect illustrated by the skewing of the RD114 virogenes toward main band density in Figure 10. This skewing probably does not arise from internal virogene G + C heterogeneity, nor

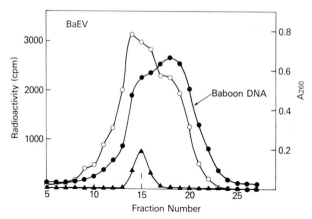

Fig. 11. Buoyant density of baboon virogenes.
Experimental procedures are described in the legend to Figure 10. 12,000 cpm of [152]I-labeled baboon endogenous virus RNA was hybridized for 20 hours to baboon DNA fractionated on NaI gradients. RNA hybridized per fraction (o—o); baboon DNA released from each filter (A_{260}) (•—•); ^{14}C-labeled *E. coli* DNA marker (cpm times $1/4$, ▲—▲)

is it a technical artifact. It arises from the fact that virogenes in cats are flanked by long domains of lower guanine + cytosine content DNA, WHEELER and GILLESPIE, in preparation, 1980, and see later).

Similar results were obtained with baboon DNA and with virogenes in baboon DNA (Fig. 11). The virogenes were detected with RNA from the baboon endogenous virus. Like cat DNA, baboon DNA banded as a heterogeneous mixture of DNA fragments with respect to buoyant density, having a median that corresponded to a guanine + cytosine content of 40%. Like the RD114 virogenes in cats, the baboon virogenes have a higher buoyant density than most of the cellular DNA, exhibiting an apparent modal G + C content of about 50%.

Virogenes from cats and baboons are related to one another; the cat virogenes originated by a germline infection of an ancestor of present day cats by baboon endogenous virus or its predecessor (BENVENISTE and TODARO, 1974 and see Section V on Transmission of Retroviruses Among Animals). The virogenes of chickens, the RAV-0 virogenes, are totally unrelated to either of the mammalian virogenes described above. However, as shown in Fig. 12, the RAV-0

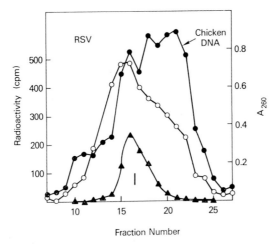

Fig. 12. Buoyant density of chicken virogenes.
Experimental procedures are described in the legend to Figure 10. Each DNA filter was prehybridized with 200 ng of unlabeled HeLa cell rRNA in 15 µl of hybridization solution. After 3 days 9,000 cpm of ^{125}I-labeled Rous sarcoma virus RNA in 20 µl of hybridization solution was added and the hybridization was continued at 37° for 6 additional days. RNA hybridized (cpm) per fraction (o—o); chicken DNA released from each filter (A_{260}) (•—•); ^{14}C-labeled *E. coli* DNA marker (cpm times $^{1}/_{2}$, ▲—▲)

virogenes, too, exhibit a guanine + cytosine content higher than that of the bulk of the DNA from the same source. The RAV-0 virogenes, as probed with RNA from Rous sarcoma virus (NEIMAN *et al.*, 1974) have a buoyant density somewhat higher then the buoyant density of the cat and baboon virogenes, in agreement with the higher guanine + cytosine content of the RNA from a closely related virus, RAV-1 (compare ROY-BURMAN and KAPLAN, 1972 with ROBINSON, PITKANEN and RUBIN, 1965 and see Table 1). Therefore, it is unlikely that the avian type-C virogene is interrupted by a long intervening sequence of low G + C content.

In addition to true virogenes the DNA of at least some animals carries regions related in nucleotide sequence to the genomes of class 2 retroviruses. The production of infectious virus from these sequences has not been observed. In cats the genes related to the feline leukemia virus are examples of this type of sequence. Fig. 13 shows that the feline leukemia virus-related genes in cats share the property of virogenes of being high $G + C$ sequences. The same is true of genes in mice related to Rauscher leukemia virus (Fig. 14) and of genes in mice related to simian sarcoma virus (Fig. 15), a virus that naturally infects primates but was thought to have arisen from genes in normal mice (BENVENISTE et al., 1974; WONG-STAAL, GALLO and GILLESPIE, 1975).

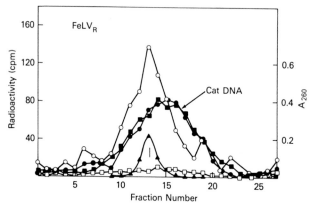

Fig. 13. Buoyant density of cat DNA related to feline leukemia virus. Experimental procedures are described in the legend to Figure 10. Each DNA filter was prehybridized with 200 ng of Hela cell rRNA in 15 µl of hybridization solution. After 2 days 1,800 cpm of ^{125}I-labeled feline leukemia virus RNA in 15 µl of hybridization solution was added and the hybridization was continued for 6 additional days. The results of two separate NaI gradients are superimposed; one shows hybridization to domestic cat DNA the other to leopard cat DNA. Leopard cats lack the feline leukemia virus-related DNA. RNA hybridized (cpm) per fraction to domestic (o—o) or leopard (□—□) cat DNA; domestic (●—●) or leopard (■—■) cat DNA released from each filter (A_{260}); ^{14}C-labeled E. coli DNA marker (cpm times 0.1, ▲—▲)

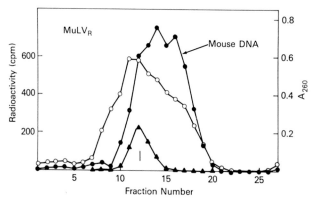

Fig. 14. Buoyant density of mouse DNA related to Rauscher leukemia virus. Experimental procedures are described in the legend to Figure 10. Each DNA filter was prehybridized with 200 ng of unlabeled HeLa cell rRNA in 15 µl of hybridization solution. After 4 days 4,500 cpm of ^{125}I-labeled Rauscher leukemia virus RNA was added and the hybridization was continued at 37° for 5 additional days. RNA hybridized (cpm) per fraction (o—o); mouse DNA released from each filter (A_{260}) (●—●); ^{14}C-labeled E. coli DNA marker (cpm times $^{1}/_{2}$, ▲—▲)

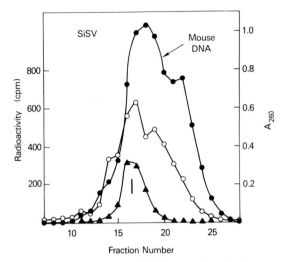

Fig. 15. Buoyant density of mouse DNA related to simian sarcoma virus. Experimental procedures are described in the legend to Figure 10. Each DNA filter was prehybridized to 200 ng of HeLa cell rRNA in 15 μl of hybridization buffer. After 3 days 16,000 cpm of ^{125}I-labeled simian sarcoma virus RNA in 20 μl of hybridization solution was added and the hybridization was continued for 6 additional days at 37°. RNA hybridized (cpm) per fraction (o—o); mouse DNA released from each filter (A$_{260}$) (●—●); ^{14}C-labeled *E. coli* DNA marker (cpm times $^1/_4$, ▲—▲)

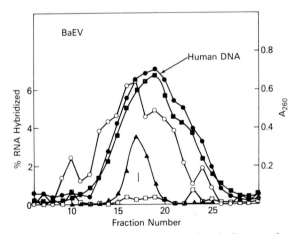

Fig. 16. Buoyant density of human DNA related to baboon endogenous virus. Experimental procedures are described in the legend to Figure 10. Each DNA filter was prehybridized with 200 ng of unlabeled HeLa cell rRNA in 15 μl of hybridization solution. After 9 days 8,000 cpm of ^{125}I-labeled baboon endogenous virus RNA in 15 μl of hybridization solution was added and the hybridization was continued for 5 additional days. The DNA from a duplicate gradient was hybridized with 9,400 cpm of ^{125}I-labeled *E. coli* rRNA (16 S + 23 S) in 20 μl of hybridization solution for 7 days. The results of the two gradients have been superimposed. The results are normalized to hybridization to unfractionated, homologous DNA. Baboon endogenous virus RNA (o—o) or *E. coli* rRNA (□—□) hybridized (% of unfractionated, homologous DNA) per fraction to human placental DNA; human DNA (●—●, ■—■); ^{14}C-labeled *E. coli* DNA marker from one gradient (▲—▲)

Therefore, virogenes are high G + C elements and have been conserved as such in birds, rodents, carnivores and primates. In a practical sense this fact adds a property of virogenes that can be used to characterize a genomic DNA sequence suspected of being viral but for which no biological assays are available. For example, there exist sequences in human DNA that are evolutionarily related to baboon virogenes. They are recognizable by a low level of hybridization of RNA or cDNA from the baboon endogenous virus to baboon DNA (BENVENISTE and TODARO, 1976; DONEHOWER, WONG-STAAL and GILLESPIE, 1977), but the hybridization is so low that "nonspecific" hybridization was not ruled out. Fig. 16 demonstrates that the human DNA sequences that hybridize to the RNA of baboon endogenous virus are a select set, being contained in DNA fragments having an apparent guanine + cytosine content of 50%. For comparison, the extent of interaction between ribosomal RNA from *E. coli*, an RNA of complexity comparable to that of the baboon endogenous virus genome and an RNA of high guanine + cytosine content is shown (open square, Fig. 16). The viral RNA-human DNA hybrid had a thermal transition in tetraethylammonium chloride (MELCHIOR and VON HIPPEL, 1973) expected of a mismatched complex (Fig. 17) and the RNA recovered from the hybrid had hybridization properties of baboon endogenous

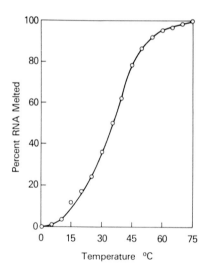

Fig. 17. Thermal melting of BaEV RNA · human DNA hybrids in 2.4 M tetraethyl-ammonium chloride (TEAC).

Ten nitrocellulose filters each containing 30 μgm of human placental DNA were prehybridized with 200 ng of unlabeled HeLa r-RNA in 15 μl of hybridization solution (50% formamide—3 × SSC). After 4 days 10,000 cpm of ^{125}I-labeled BaEV RNA in 15 μl was added and the hybridization was continued at 37° C for 4 additional days. After exhaustive washing with 3 × SSC the filters were incubated in 2.4 M TEAC at increasing temperatures (0° → 75° C at 5° C intervals) for 5 minutes at each temperature. The ^{125}I-labeled BaEV RNA released at each temperature was measured in a gamma counter. The results obtained for a typical filter are shown above. The BaEV RNA released between 20°—55° C from 10 filters was pooled, precipitated with 2 volumes of ethanol, recovered by centrifugation at 10,000 × g for 2 hours and hybridized to baboon or dog DNA immobilized on nitrocellulose filters (see text for results)

virus RNA. On the other hand, there has been no report of the isolation of an endogenous virus from humans, so the hybridization results can only be classified as "consistent" with humans having type-C virogenes.

In a conceptual sense the conservation of guanine + cytosine content among vertebrate virogenes is puzzling. Endogenous viral sequences divergent for more than 300 million years have nearly the same guanine + cytosine content. This is despite the unusually rapid evolution of virogenes relative to most unique DNA sequences (GILLESPIE et al., 1975; BENVENISTE and TODARO, 1976; DONEHOWER et al., 1977). This rapid evolution is thought to arise in part from repeated recombination events during infection in addition to rapid accumulation of tolerated mutations (ALTANER and TEMIN, 1970; CALLAHAN et al., 1974). Thus, among type-C viruses, the conservation of guanine + cytosine content does not result from a conservation of base sequence. In particular, since many of the viruses examined lack detectable protein homology, conservation of guanine + cytosine content does not result from conservation of coding sequences.

Some clue concerning the reasons for conserving the guanine + cytosine content of type-C viral genomes might be obtained from examining base compositional features of the RNA transcripts of virogenes, the viral genome. Table 1 shows the best estimates of base composition of the genomes of nine retroviruses along with the means and variances (s^2) for each individual base and for three combinations of two bases. The $G + C$ combination shows the narrowest range of values among the nine viruses with a variance of 1.70 which is even lower than the variance of any individual base. The variation of s^2 between the three combination of pairs is greater than expected by chance alone with $p < 0.025$[1]. In addition, the s^2 of $G + C$ is significantly lower statistically than the s^2 of $C + A$ ($p < .006$) or $C + U$ ($p < .05$). Thus, guanine+cytosine (or adenine plus thymine) content is the most conserved base compositional feature among retroviral RNAs. The variation of s^2 between the four bases themselves is not statistically significant with $p > 0.80$. Therefore, conservation of $G + C$ does not merely reflect a conservation of G and C independently. Instead, as G (or C) increases, C (or G) decreases to keep the sum $(G + C)$ constant.

[1] The homogeneity of the variances in Table 1 was evaluated by using

$$\frac{-2\ln\mu}{1 + \frac{1}{3(k-1)}\left(\sum_{i=1}^{k}\frac{1}{n_i-1} - \frac{1}{n-k}\right)}$$

with $\mu = \dfrac{\prod_{i=1}^{k}\left(\dfrac{n_i s_i^2}{n_i-1}\right)^{\frac{n_i-1}{2}}}{\left(\dfrac{\Sigma n_i s_i^2}{\Sigma(n_i-1)}\right)^{\frac{\Sigma n_i-1}{2}}}$

as a variable having a χ^2 distribution with $k-1$ degrees of freedom (HOEL, 1962). The variances $s_1, s_2, \ldots, s_k$ to be evaluated have sample sizes $n_1, n_2, \ldots, n_k$ with

$$n = \sum_{i=1}^{k} n_i.$$

Although the conservation of viral RNA structure for the purpose of packaging in virions may be important and may account for conserved features in the structure of virogenes, we were unable to account for the guanine plus cytosine conservation at the RNA level. It seems unlikely that the regulation of guanine—cytosine content could relate to RNA primary structure for the reasons given above or to the secondary structure of the viral RNA since guanine equals cytosine would seem a more likely parameter to be conserved.

In DNA the regulation of guanine plus cytosine content results in regulation of interstrand stability (MARMUR and DOTY, 1959). Regions of homogeneous physical stability have been indicated in phage DNA (WADA et al., 1976). Such homostability regions may play important roles in transcriptional control. For example, the ease of DNA unwinding could determine the quantity of a particular gene transcript in the absence of discriminating initiation signals.

Alternatively, the conservation of guanine plus cytosine content may relate to the organization of proteins on DNA. It has been proposed that β- ribbon portions of proteins interact with the minor groove of RNA (CARTER and KRAUT,

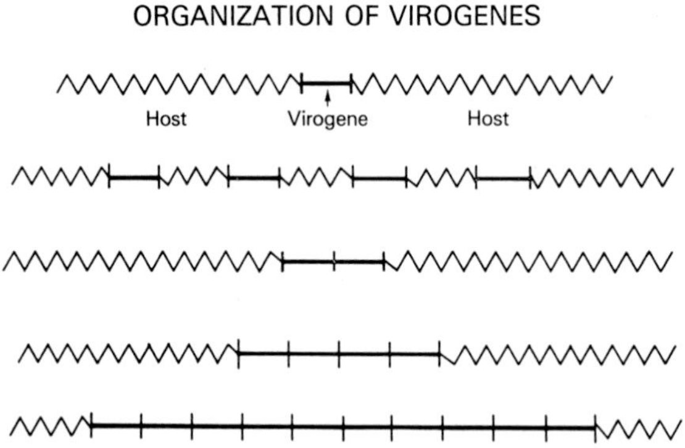

ORGANIZATION OF VIROGENES

Fig. 18. Diagram of possible organizations of virogenes

1974), or DNA duplexes (CHURCH, SUSSMAN and KIM, 1977). This is a base sequence-independent interaction but the protein is free to slide in the DNA minor groove and probe with hydrogen-bonding groups for base-specific contacts. SEEMAN, ROSENBERG and RICH (1976) proposed a system that utilizes two H-bonding sites for the recognition of each base pair in double stranded DNA. With respect to protein probes the A—T and T—A base pairs are equivalent in the DNA minor groove, while essentially all other base pairs exhibit unique stereo-chemistry. If the arrangement of AT pairs regulated the interaction of control proteins on DNA, then the involvement of a significant number of base pairs in the recognition process would both fix the AT content of a region of DNA and could contribute to its nucleoprotein structure. This would result in the apparent conservation of guanine—cytosine content.

By examining the buoyant density of larger DNA fragments, it is possible to show that the length of the 50% G + C region is about 10 kilobase pairs and that these regions are flanked by large domains (> 25 kbp) of lower guanine + cytosine content. The general logic of this experiment is depicted in Fig. 18. If virogenes are very long stretches of high G + C content DNA, for example, if they are arranged in tandem, then most randomly-sheared DNA fragments up to 50 kilobase pairs in length which contain virogene sequences will still have a guanine + cytosine content of 50%. Conversely, if virogenes are organized as separated genes flanked by long stretches of DNA with an average guanine + cytosine content of 40%, then randomly-sheared DNA fragments as short as 20 kilobase pairs will, on the average, be composed equally of 50% G + C virogene and 40% G + C nonvirogene sequences. The actual curves that relate density of virogene-containing DNA fragments to fragment-chain length will be characteristic for each organization. Some theoretical curves are shown in Fig. 19. They were calculated

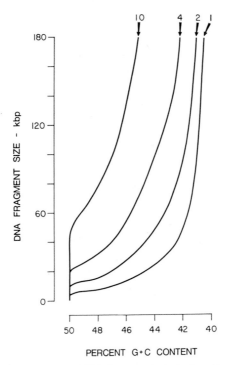

Fig. 19. Theoretical curve relating DNA fragment chain length and virogene density. See text for explanation

on the assumption that a virogene contains 10 kilobase pairs, that it has a 50% guanine + cytosine content throughout its length and that flanking cell sequences have a guanine + cytosine content of 40%, like most cell DNA sequences. A model having two tandem virogenes of 50% guanine + cytosine content is formally the same as one virogene carrying 10 kilobase pairs of high G + C content intervening sequences. Thus, finding the special case of individual organization (top model, Fig. 18) rules out high G + C intervening sequences of large size, (i.e. > 5 kilobase pairs), as well as indicates a separated organization.

To differentiate among the models shown in Fig. 18 with respect to the RD 114 virogenes in cats, cat DNA was randomly sheared then separated according to size by electrophoresis through gels of agarose on a preparative scale (STRAYER,

WHEELER and GILLESPIE, in preparation). DNA was removed from the agarose gel by centrifugation in NaI, then the buoyant density of DNA fragments containing virogenes was measured for different size classes of DNA. Fig. 20 presents the size distribution of the DNA recovered from the agarose prep gel and after the

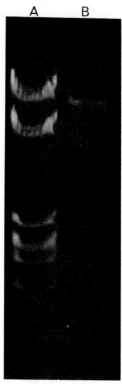

Fig. 20. Size estimation of DNA fragments analyzed in Fig. 21.
Randomly sheared cat DNA was fractionated according to size by electrophoresis through 0.4% agarose on a preparative scale (see legend to Fig. 25). DNA was recovered from the agarose gel by centrifugation to equilibrium after dissolving the agarose in NaI. The DNA was banded analytically to measure the buoyant of the RD114 virogenes. Before the DNA was denatured and immobilized on nitrocellulose an aliquot was taken for sizing on a slab of 0.3% agarose. Electrophoresis was in 0.04 M Tris-acetate, pH 7.8, 0.005 M Na acetate and 0.002 M EDTA for 20 hours at 8 ma. The gel was stained in 1 μg/ml of ethidium bromide for 60 minutes, then photographed under UV light. Slot A = lambda virus DNA plus lambda virus DNA digested with Eco R 1 endonuclease. Slot B = 28 million dalton cat DNA (46 kilobase pairs)

final centrifugation in NaI. Fig. 21 shows the density distribution of the DNA fragments 46 kilobase pairs long and containing virogene sequences. Virogenes contained in DNA fragments of this large size are of much lower density than virogenes contained in DNA 5 kilobase pairs in length (compare Figs. 10 and 21). The density at the center of the virogene band of 46 kilobase pair DNA corresponds to a guanine + cytosine content of 39.7%, yielding a point that best fits the computer-generated curve of an organization based on individually-located viro-

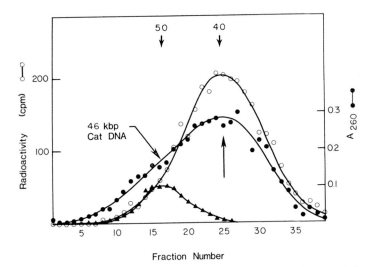

Fraction Number

Fig. 21. Buoyant density of high molecular weight cat DNA containing RD114 virogenes.

DNA fractionated as described in the legend to Figure 20 was mixed with *E. coli* DNA marker and centrifuged at 23,000 rpm for 250 hours. Other experimental procedures are those described in the legend to Figure 10. Each DNA filter was prehybridized with 200 ng of unlabeled Hela cell rRNA in 15 μl of hybridization solution. After 3 days 2,000 cpm of ^{125}I-labeled RD114 viral RNA was added and hybridization was continued for an additional 5 days. RNA hybridized (cpm) per fraction (o—o); cat DNA released from each filter (A_{260}) (•—•); ^{14}C-labeled *E. coli* DNA marker (cpm times 10) (▲—▲)

genes in the cat genome (Fig. 22 A). Experiments were performed to study additional size classes of cat DNA and the results shown in Figs. 22 A and B suggest that each RD114 virogene (10 kilobase pairs) has a high G + C content (*e.g.* 50%) and is flanked by regions of lower G + C content (*e.g.* 38%). The simplest interpretation at this point seems to be not only that virogenes are separated from one another, but also that they are not internally interrupted by large high G + C (>5 kbp) or low G + C (>2 kbp) intervening DNA. We cannot rule out the possibility that virogenes are both fractured *and* carry intervening sequences of high G + C content, but we consider this to be unlikely. We also cannot exclude the possibility that virogene units are in tandem but consist of a much larger domain than is indicated by the length of the retroviral RNA, *i.e.* virogenes are separated by spacers of low G + C content DNA which are at least 25 kbp in length.

There are two *a priori* reasons for expecting a tandem organization of virogenes, especially in the cat. The cat RD114 virogenes were newly introduced into the species, at least in evolutionary terms. They apparently entered cats by infection of germline cells some 3—5 million years ago (BENVENISTE and TODARO, 1974). If they entered as single-copy DNA, then they were amplified into 10-copy genes. The mechanisms of amplification proposed and known to us—involving replicative loops or unequal crossing over (KEYL, 1965; SMITH, 1973 and 1976)—result in tandemly organized amplification products.

Second, some multiple-copy genes *are* tandemly organized. Ribosomal DNA and histone genes (KEDES *et al.*, 1975) display a tandem organization.

A first indication that virogenes are tandemly-arranged, but separated by low G + C spacers could come from studies of the regions that immediately flank them. If each of the approximately ten virogenes in an animal has the same flanking

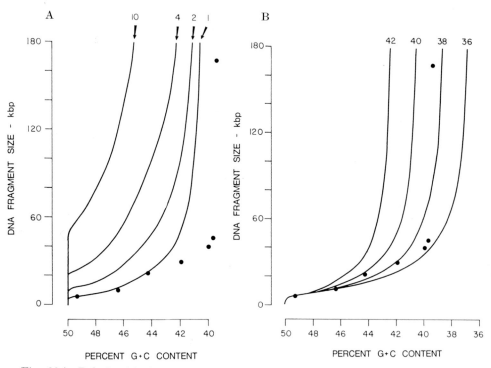

Fig. 22A. Relationship between length and G + C content of cat DNA fragments containing RD 114 virogenes.

Distinguishing between tandem and individual organizations. Cat DNA was randomly sheared, fractionated according to size and the buoyant density of DNA fragments containing RD 114 virogenes in different DNA size classes was determined as detailed in the legends to Figs. 10, 20 and 21. The results have been plotted on the theoretical curves described in the text (Fig. 19)

Fig. 22B. Relationship between length and G + C content of cat DNA fragments containing RD 114 virogenes.

Distinguishing between several G + C contents for flanking sequences. The same experimental results (Fig. 22A) are shown with four theoretical curves. These curves assume an individual virogene organization, virogene G + C content = 50% and virogene size = 10 kbp. The G + C content of flanking sequences was varied (36, 38 40 and 42%)

sequence, then virogenes could reside within domains of low guanine + cytosine content, apparently individually organized, but still be part of a tandem block. The following experiments involve studies of baboon virogenes and do indicate such conservation of flanking regions.

Like the RD114 virogenes of cats, each baboon virogene is also flanked by domains of low guanine + cytosine content (Figs. 11, 23 and 24). Fig. 23 demonstrates that 140 kilobase pairs-long fragments of baboon DNA containing virogene sequences are lower density than virogene containing fragments 5.5 kilobase pairs long (Fig. 11). The density relationship of virogene-containing DNA vs.

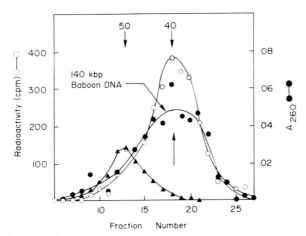

Fig. 23. Buoyant density of high molecular weight baboon DNA containing baboon endogenous virogenes.
Baboon DNA (140 kbp) obtained as described in the legend to Figure 20 was mixed with *E. coli* DNA marker and centrifuged at 27,000 rpm for 200 hours. Other experimental procedures are those described in the legend to Figure 10. Each DNA filter was prehybridized with 200 ng of unlabeled HeLa cell rRNA in 15 µl of hybridization solution. After 2 days 2,000 cpm of ^{125}I-labeled baboon endogenous virus RNA was added and hybridization was continued for an additional 5 days. RNA hybridized (cpm) per fraction (o—o); baboon DNA released from each filter (A$_{260}$) (•—•); ^{14}C-labeled *E. coli* DNA marker (cpm times 10) (▲—▲)

DNA fragment length is that expected of a virogene organization consisting of 10 kilobase pairs of uninterrupted DNA of 50% guanine + cytosine content flanked by a long region of about 38% G + C content (Figs. 24 A and B).

Virogenes in baboons are evolutionarily old genes, since related DNA sequences can be found by molecular hybridization in DNA of all old world primates, including apes and humans (BENVENISTE and TODARO, 1976; DONEHOWER, WONG-STAAL and GILLESPIE, 1977). Old world monkeys and apes shared a common ancestor some 30—50 million years ago, so the baboon virogenes must be at least that old. DONEHOWER, WONG-STAAL and GILLESPIE (1977) postulated from RNA-DNA hybridization data that the multiple copies of baboon virogenes were nonidentical and that different regions of the virogenes evolved at different rates. This conclusion is supported by the analysis of fragments of baboon virogenes produced by restriction endonucleases (Fig. 25). The restriction endonuclease, Endo · Eco Rl, cuts DNA at the sequence GAATTC, cutting DNA once every 5,000 bases on the average. It would be expected to cut the 10 kilobase pair baboon virogene about twice, leaving three DNA fragments containing virogene sequences, one internal fragment and two terminal fragments bearing

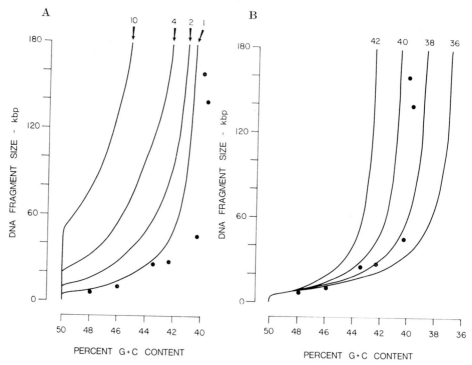

Fig. 24 A. Relationship between length and G + C content of baboon DNA fragments containing BaEV virogenes.
Distinguishing between tandem and individual organizations. Baboon DNA was randomly sheared, fractionated according to size and the buoyant density of DNA fragments containing baboon endogenous virogenes in different DNA size classes was determined as described in the legends to Figs. 10 and 23. The results have been plotted on the theoretical curves described in the text (Fig. 19) Fig. 24 B. Relationship between length and G + C content of baboon DNA fragments containing BaEV virogenes
Distinguishing between several G + C contents for flanking sequences. See legend Fig. 22 B

flanking cell DNA. In practice when DNA is cut with Endo · Eco Rl and the resulting fragments are separated according to size on agarose gels many fragments are found, so many that a distribution of sequences is generated approximating the average size of the total population of baboon DNA fragments with a few peaks rising out of the population (Fig. 25). The two major peaks are 8.1 kilobase pairs and 4.6 kilobase pairs, judging from quantitative analysis of the virogene concentration in each gel slice (see legend to Fig. 25).

Knowing that baboon virogenes are individually-organized genes of 50% G + C content flanked by lower G + C DNA and appreciating their sequence heterogeneity with respect to fragmentation by Endo · Eco Rl, it becomes possible to assess the homogeneity or heterogeneity of the flanking sequences. If the flanking sequences are heterogeneous, then the distance between the virogene terminus and the first Endo · Eco Rl cleavage site in the flanking sequence will not be preserved from one virogene to the next. Internal fragments will tend to be more

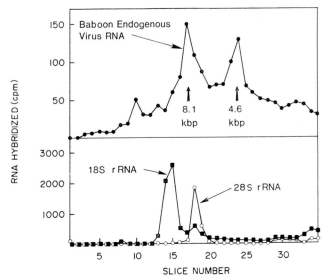

Fig. 25. Size of virogene-containing fragments in baboon DNA digested with Eco R 1 endonuclease.

Twelve milligrams of baboon DNA was digested with 12,000 units of Endo · Eco R 1 at 37° for 6 hours. Completion of digestion and lack of random degradation was monitored by examining the nature of ribosomal DNA fragments (lower panel). The DNA was made 10% in glycerol and applied to the surface of an 0.6% agarose cylinder 5 inches in diameter and 7 inches in length. Electrophoresis was at 4° and 12 volts for 7 days. After electrophoresis the gel was sectioned into 0.4 cm slices. The size distribution of DNA in each slice was determined by placing a small piece of the gel on a slab gel and electrophoresing the DNA through the slab in the presence of suitable markers. The concentration of baboon endogenous virogenes in each gel slice was measured by hybridization in DNA excess as described by VOGELSTEIN and GILLESPIE (1979). Briefly, a representative portion of each slice was dissolved in NaI and DNA was selectively precipitated from the solution with 0.5 vols of acetone. The DNA was collected by centrifugation and overlaid with 10 μl of hybridization solution (70% formamide and 0.4 M phosphate; VOGELSTEIN and GILLESPIE, 1977) containing ^{125}I-labeled RNA from baboon endogenous virus. The system was overlaid with mineral oil, incubated at 100° for 10 minutes then at 44° for 3 days. Hybrids were detected by resistance to ribonuclease. Similarly, slices containing 18S and 28S rDNA were located by molecular hybridization

homogeneous in size than terminal fragments (Fig. 26). Thus, both the 8.1 kilobase pair and the 4.6 kilobase pair Endo · Eco Rl virogene DNA fragments will be internal, representing two divergent forms of the baboon virogenes. If either fragment were a terminal fragment, the fact would support the idea that virogene flanking regions are conserved. Terminal and internal virogene fragments can be resolved by centrifugation to equilibrium in NaI gradients. It is important to note that the experiment involves *specific* DNA fragments, produced by nucleases that cleave DNA at specific sites and is not constrained by the same logic used to assess density distributions of randomly-sheared DNA. Thus, internal fragments will lack low G + C flanking domains and will have a higher buoyant density than terminal virogene fragments. The density of terminal fragments will be determined by the ratio of virogene: flanking DNA.

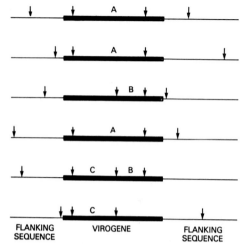

Fig. 26. Size distributions of endonuclease-treated virogenes with nonconserved flanking sequences

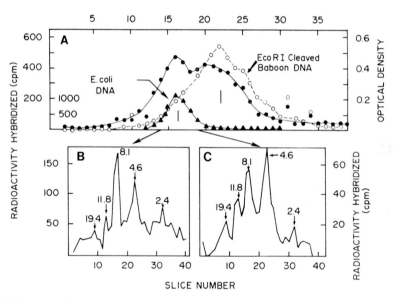

Fig. 27. Buoyant density of virogenes in baboon DNA digested with Eco R 1 and size of high and low density fragments.

Baboon DNA was digested with Endo · Eco R 1 and fractionated according to buoyant density as described in the legend to Fig. 10. Fractions containing baboon virogenes were located by hybridization to ^{125}I-labeled RNA from baboon endogenous virus. A: DNA from fractions 14 and 19 were precipitated from ethanol and fractionated by electrophoresis through gels of 0.6% agarose. The gel was sliced and virogenes were located by acetone precipitation and molecular hybridization as outlined in the legend to Fig. 25. B: Gel electrophoresis of DNA from NaI gradient fraction 14; C: Gel electrophoresis of DNA from NaI gradient fraction 19. Numbers above peaks represent size of DNA fragments in kilobase pairs

It should be possible to resolve terminal from internal fragments by measuring the size of fragments first separated according to buoyant density or by measuring the density of fragments first separated according to size. Fig. 27 A shows the density profile of baboon virogenes that have been cut specifically with Endo · Eco Rl. There appear to be two density distributions; one centered around 50% G + C and one with a lower average buoyant density. Figs. 27 B and 27 C show that the predominant fragments present in both areas of the gradient are the 4.6 and the 8.1 kilobase pair fragments. The method is not reliable enough to indicate whether either fragment is terminal or internal, since each density cut contained both fragments.

Figs. 25 and 28 show the reverse experiment. Baboon DNA treated with Endo · Eco Rl was fractionated according to size by electrophoresis through

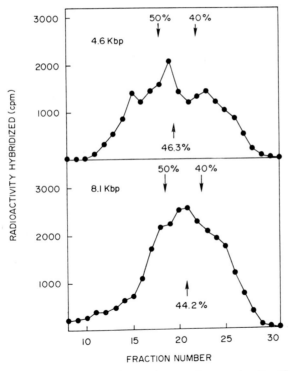

Fig. 28. Buoyant density of major baboon virogene fragments after Eco R 1 cleavage. The 8.1 and 4.6 kbp DNA fragments (slices 17 and 24, Fig. 25) were recovered from the agarose gel as described in the legend to Fig. 20. The buoyant density of each purified fragment was studied in gradients of NaI as described previously (legend to Fig. 10). ^{14}C-labeled *E. coli* DNA (50% G + C) and ^{3}H-labeled SV 40 DNA (40% were included in each gradient to serve as density markers. ^{125}I-labeled baboon endogenous virus RNA was hybridized to each gradient fraction as described in the legend to Fig. 10. The results of duplicate experiments are shown in Table 2

agarose (Fig. 25) and the 4.6 and 8.1 kilobase pair fragments were isolated. Their apparent $G + C$ content was measured by centrifugation to equilibrium in gradients of NaI (Fig. 28). The buoyant density of the 4.6 and 8.1 kilobase pair fragments corresponded to $G + C$ contents of 46.7 and 44.4, respectively. Duplicate measurements yielded similar results (Table 2). Thus, *both* fragments are probably

Table 2. *Eco RI fragments of baboon virogenes*

DNA fragment size	G + C content (%)		
kbp	Exp. 1	Exp. 2	Average
4.6	47.1	46.3	46.7 ± 0.4
8.1	44.5	44.2	44.4 ± 0.2

terminal, although, we cannot rule out the possibility that a 2 kilobase pair (upper size limit permitted from our data) intervening sequence of low $G + C$ content accounts for the buoyant density of the 4.6 kilobase pair fragment. This result suggests that the areas surrounding the 10 kilobase pair virogene are conserved. Therefore, multiplication of virogenes from a single copy gene to a 10-copy family of genes probably did not arise by interspersion to unselected chromosome locations.

It would be interesting to know what virogenes do in their normal setting. They are expressed during differentiation (STOCKERT, OLD and BOYCE, 1971; RISSER *et al.*, 1978). It seems unreasonable to relegate them to the role of generating viruses, especially considering that the viruses, even the endogenous ones, have the potential for becoming oncogenic. Possibly, they have a role making use of a typical property of viruses, *e.g.* that of transmission, without requiring that they fulfill *all* the criteria that define a virus, *i.e.* an autonomous genetic existence. MAYER, SMITH and GALLO proposed in 1974 that the baboon endogenous virus might effect transfer of information from the mother to her offspring during embryogenesis. It is difficult to reconcile this with the genetic similarity between mother and child, but perhaps the virogenes operate by gene relocation or rearrangement within a cell, rather than between cells and individuals. The genes may be normally quiescent unless activated in certain cells by relocation, possibly using the mechanism described by TEMIN's protovirus theory (1971). Probably, virogenes would be more prone to interanimal transmission while they are being activated for relocation.

Since virogenes code for endogenous viruses and endogenous viruses can be precursors to tumorigenic RNA viruses, virogenes are at least potentially tumorigenic themselves. Indeed, HUEBNER and TODARO in 1969 proposed that cancer is the "turned on" state of virogenes and an associated gene, the oncogene. Since TEMIN proposed his proto-virus model in 1970, a model that demanded genetic change of virogene products to affect carcinogenesis, it has become unfashionable to seriously consider the virogene-oncogene model. We know that the oncogenic sequences of an RNA tumor virus are inherited from cell genes (STEHELIN et al., 1976) however, it is questionable whether src or its cellular progenitor, protosrc is preferentially expressed in tumors (BONDURANT et al., 1979). Other genes appear to be preferentially transcribed in transformed cells (GROUDINE and WEINTRAUB, 1975; ADAMS et al., 1977). The observations that chicken cells produce mRNA coding for fetal hemoglobin after infection by Rous sarcoma virus (GROUDINE and WEINTRAUB, 1975) might suggest a special relationship between type-C viruses and hematopoietic functions, but the work of ADAMS et al. (1977) suggests that the expression of several genes is altered after infection. In any event it may be reasonable to think in terms of the deregulation of cell genes by an incoming virus or by a chemical mutagen.

One interaction of special interest is that between an incoming virus genome and cellular virogenes. Do proviruses integrate within or next to virogenes? Can exogenously added viral genes modulate the expression of virogenes? Does this modulation affect cellular growth patterns? RISSER, STOCKERT and OLD (1978) report that an antigen peculiar to Abelson murine leukemia virus is also found on normal cells from marrow, spleen and fetal liver but not on cells from several adult tissues. At the end of the discussion of their work they wonder ". . . what effects on normal cell function and proliferative state such a molecule might have if it were constitutively expressed as a consequence of viral infection". This idea, that constitutive expression of virogenes, may impair differentiation, cause "de-differentiation", or otherwise alter growth patterns in producing a "transformed" state seems to be becoming more popular. In such a context, the RNA tumor viruses can be compared to DNA viruses and carcinogens. The viruses can be viewed as site-specific mutagens or activators because of their sequence homology with virogenes. Mutagenic carcinogens could disrupt gene expression more generally by nucleotide sequence alternation at random positions on the chromo-some. Nonmutagenic carcinogens might alter virogene expression at a post-transcriptional level, albeit some mechanism for establishing long-term alterations in a pluripotent cell must be proposed in this case.

In any event, it is important to consider the idea that modulation of the regulation of virogenes by infecting RNA tumor viruses can cause cell trans-formation. The sections to follow on the transmission of Retroviruses among cells and on the relatedness among Retroviruses (especially recombinant viruses) explore this possibility more deeply.

IV. Organization of Infectious Retrovirus Genes

Transmission of RNA tumor viruses begins when virogenes are expressed and the gene products are packaged in an extracellular particle. The virus released usually grows more readily in cells from a new animal species (secondary cells) than in cells from the species that released them. The *capacity* to produce virus is transmitted vertically by both the first and second hosts. Expression of type-C virus information represents a spectrum of biological activities; second-arily-infected cells can produce biologically active virus, biologically inactive particles or produce no particles even though infected.

Cells infected by an RNA tumor virus contain a DNA provirus integrated into the infected cell genome, hence the capacity to produce the viruses can be inher-ited. Although TEMIN proposed the provirus model in 1964, it gathered few supporters until 1970 when he and MIZUTANI (1970) and BALTIMORE (1970) discovered an enzyme capable of making the requisite DNA copies of the viral RNA genome. Scientific opinion shifted dramatically until TEMIN's original idea, in its simplest possible form was added to the central dogma. In a way this has been unfortunate because while the provirus concept is suitable for explaining transmission of retroviruses among cells, those natural infections that result in oncogenesis may be more complicated. Indeed, TEMIN (1971) in formulating his protovirus hypothesis (see Section II on Origin of Retroviruses) indicated that a more complex process was likely.

In its simplest form the provirus theory states: (1) viral RNA codes for viral functions (see Figs. 4 and 5), (2) during infection a DNA copy of the viral RNA is made, (3) this "free provirus" is integrated *in toto* into the chromosome of the host cell, and (4) the "integrated provirus" is capable of being expressed in a manner that leads to production of the original virus by the new host cell (Fig. 4). Aside from many studies on the detailed mechanism of action of reverse trans-criptase, the provirus model is supported by molecular hybridization and DNA transfection experiments.

TEMIN (1964) and BALUDA (BALUDA and NAYAK, 1970; BALUDA, 1972) pioneered the molecular hybridization studies using avian RNA tumor viruses. Using an excess of RNA purified from RNA tumor viruses they were able to show new viral-related sequences in the DNA of virus-infected cells that were not detectable in the cells prior to infection. These experiments had three dif-ficulties that limited interpretation. First, RNA tumor viruses contain cell components since they bud from the cell membrane. Particularly, impure virus preparations contain cell-derived nonviral RNA molecules. Though viral RNA preparations free of cell contaminants were prepared occasionally (BADER and STECK, 1969) this was uncommon. Consequently, BALUDA and TEMIN carried out their molecular hybridization experiments in the presence of an excess of unlabeled (competing) cellular RNA. We know now that many of the sequences in RNA tumor virus genomes are mirrored by related cell RNA and DNA, so in these experiments some of the viral sequences were probably competed as well. Moreover, we cannot know the efficiency of competition of the cell-derived RNA. Second, the uninfected cells carried DNA sequences related to a portion of the RNA genomes of the viruses studied. Thus, new sequences in the infected cell was reflected by about two times higher hybridization values with DNA from the infected cell than with DNA from the uninfected cell. Third, at the time BALUDA and TEMIN did their early work the Carnegie group was just beginning to learn about repetition of DNA sequences in the genomes of animals (BRITTEN and WARING, 1965) and the complications that sequence repetition has on interpreting molecular hybridization experiments (BRITTEN and KOHNE, 1968). TEMIN and BALUDA could not have known the repetition frequency of those DNA sequences of the infected cell that became involved in hybrids with viral RNA. If they were repeated sequences, then the conclusion that similar sequences were *missing* in the uninfected cell was not warranted, they could only say that the sequences were not detectable. Indeed, more recent molecular hybridization work from BALUDA's laboratory on proviral sequences in DNA of infected chicken cells shows hybridization of the viral RNA to a repeated DNA sequence component (DROHAN *et al.*, 1975).

Both TEMIN and BALUDA interpreted their results as support of the provirus theory, instead of reflecting other biological phenomena such as gene amplification or as indicating technical difficulties such as differential extraction of particular DNA sequences and, in fact, subsequent, more detailed work supports their interpretation.

Since these early molecular hybridization experiments, several investigators confirmed the presence of new, viral-related sequences in DNA from cells infected by class 2 viruses (or heterologous hosts infected by class 1 viruses) that are missing in uninfected cells. The appearance of these proviral sequences follows infection by all studied RNA tumor viruses, including the chicken viruses, Rous sarcoma virus (TEMIN, 1964; NEIMAN, 1972) and avian myeloblastosis virus (BALUDA and NAYAK, 1970); other bird viruses, the Trager duck necrosis virus (KANG and TEMIN, 1974) and duck reticuloendotheliosis virus (KANG and TEMIN, 1974; COLLET, KIERAS and FARAS, 1975); the feline viruses, feline leukemia virus (Table 2) and RD114 virus (RUPRECHT *et al.*, 1973); the mouse leukemia (GILLESPIE *et al.*, 1973; CHATTOPADHYAY *et al.*, 1974) and sarcoma viruses (GILLESPIE, 1973; SCOLNICK *et al.*, 1974); bovine leukemia virus (CALLAHAN *et al.*, 1976; KETTMAN *et al.*, 1976) and the primate viruses, baboon endogenous virus (BENVENISTE and TODARO, 1974), Mason-Pfizer monkey virus (DROHAN *et al.*, 1977), simian (woolly monkey) sarcoma virus (SCOLNICK *et al.*, 1974) and gibbon ape leukemia virus (SCOLNICK *et al.*, 1974). Both class 1 and class 2 viruses insert proviral genes into new host cells upon infection. No exceptions to this finding have been reported, though in some instances proviral DNA is more difficult to detect than would have been expected (see later). These experiments have all been done with cell DNA in vast excess, using BRITTEN and KOHNE's paper (1968) on repeated DNA sequences as a guide. They have uniformly failed to find exact class 2 viral genes in normal cells at a level of one copy per genome. Thus, it does not appear that uninfected cells have the capacity to code for the viruses studied and models such as amplification of chromosomal genes do not explain the early results of BALUDA and TEMIN.

In many instances class 2 viruses growing in cells of their natural host have been studied. The results of hybridization of RNA from class 2 viruses to DNA from natural host cells usually take the form of a low level of hybridization to DNA from uninfected cells, usually involving 10—30% of the RNA and a larger fraction of the RNA hybridized to DNA from virus-infected cells. The hybrids that are formed with DNA from uninfected cells involve DNA repeated some 10—100 times and are poorly base-paired (GILLESPIE *et al.*, 1973; BENVENISTE and TODARO, 1974; GILLESPIE and GALLO, 1975). These hybrids probably involve viral RNA sequences that have maintained some homology with host virogenes by repeated recombination (GILLESPIE, GILLESPIE and WONG-STAAL, 1975). In some instances involving transfer of a virus between species, relatedness to genes in the original species reflect vestigial homology as the virus moves genetically away from its old host (see Section VI on Relatedness Among Retroviruses).

In the case of class 2 viruses that have remained within the progenitor host virtually all of the viral RNA can be distantly related to the uninfected cell genome, since nearly all of the viral RNA can be hybridized when the formation of very poorly paired hybrids is permitted (GILLESPIE, GILLESPIE and WONG-STAAL, 1975). Of course, when DNA from the infected natural host is used well-paired hybrids involving essentially the entire virus genome can be formed.

The clearest results concerning the relatedness of a viral genome to DNA of its natural host has come from molecular hybridization experiments employing viral RNA, not cDNA reverse transcripts of the viral RNA. Even in cases where it is clear from viral RNA-cell DNA hybridization that a virus possesses novel RNA sequences not found in the natural host, cDNA-DNA hybridizations may show complete homology of the viral probe with cell DNA sequences. (Compare VARMUS et al., 1972 with NEIMAN, 1972 for analysis of Rous sarcoma virus; compare GELB, AARONSON and MARTIN, 1971 with GILLESPIE, GILLESPIE and WONG-STAAL, 1975 for analysis of Rauscher leukemia virus; and compare OKABE et al., 1976 with KOSHY et al., 1979, for analysis of feline leukemia virus.)

Feline leukemia virus is an important example of the limitations of cDNA probes when searching for virogenes or viral-related genes in cell DNA. Feline leukemia virus is horizontally transmitted among cats (ESSEX et al., 1971; HARDY et al., 1973; JARRETT et al., 1973) and is the etiologic agent for most cat leukemias and lymphomas (BRODY et al., 1969). Evidence suggests that the virus was transmitted from rats to cats about 3—5 million years ago (BENVENISTE, SHERR and TODARO, 1975, and see following section), almost exactly the same time that RD 114 was transmitted from baboons to cats. RD 114 infected the cat germline and became a *bona fide* cat virogene. RD 114 seems not to cause cat leukemia.

The situation with feline leukemia virus is less clear. From cDNA hybridization data (BENVENISTE and TODARO, 1977; OKABE et al., 1976) it appears endogenous, but from viral RNA hybridization results it is clearly a class 2 virus possessing genes not found as such in DNA of normal cats (KOSHY et al., 1979). Thus, cat leukemia is associated with the introduction of a foreign genetic element carried by a virus with genetic homology with its host, just as in the cases of mice and chickens.

Occasionally cDNA transcripts can be used to define novel sequences in a virus. Host-related sequences can be removed by preparative hybridization (BAXT and SPIEGELMAN, 1972; RUPRECHT et al., 1973). Sometimes, the novel sequences can be seen by paying extreme attention to detail, as in the case of the AKR mouse leukemia virus (CHATTOPADHYAY et al., 1974; LOWY et al., 1974).

The second type of experiment supporting the provirus model is the DNA transfection experiment. First accomplished by HILL and HILLOVA (1972) and since confirmed by several investigators, the experiment consists of extracting DNA from virus-infected cells then using this DNA to "transfect" normal cells, causing them to produce the same class 2 virus produced by the original infected cells used to make the DNA. Like the experiments resulting in induction of type-C RNA viruses from normal cells, which confirmed the virogene theory at a functional level, DNA transfection confirms the provirus model at a functional level. Work through 1975—1976 leaves little doubt that the provirus concept is essentially correct. More recent work defines the mechanism of provirus synthesis, including the cytoplasmic (VARMUS et al., 1978) synthesis of supercoiled viral DNA (GUN-TAKA et al., 1975) and its migration to the nucleus prior to recombination with the host chromosome (SHANK and VARMUS, 1978).

It is generally accepted that free provirus recombines with a host chromosome and becomes a part of it. However the fact that proviral DNA is physically integrated into the host chromosome as linear cell-viral-cell DNA is poorly documented. BISHOP, VARMUS and their colleagues showed that, when DNA of infected cells was denatured then reannealed to low C_0t, the resultant DNA network contains viral sequences (VARMUS, BISHOP and VOGT, 1973). Networks can form upon reannealing long pieces of DNA because each molecule carries several repeated DNA sequences and highly branched structures can be formed. The network is essentially insoluble, while unreacted DNA remains in solution. Since viral sequences were found in the insoluble fraction, it was concluded that they were physically adjacent to or at least near cell repeated sequences, hence integrated. Of course, if the viral genome itself carried a cell repeated DNA sequence it would associate itself with the network, whether or not it was integrated. And viral genomes probably do carry cell repeated DNA sequences (DROHAN et al., 1975, and see later).

MARKHAM and BALUDA (1973) showed that the viral DNA sequences of the infected cell were carried on DNA molecules larger than the viral genome and concluded that the viral genome must therefore be integrated. But an episomal concatomer of viral genes would also be larger than the original viral DNA. The sequences adjacent to the viral genes were unfortunately not identified as cell DNA in this study.

The finding that recombinant viruses can be formed after infection (SCOLNICK et al., 1973), especially the type that appear to be recombinant between provirus and virogene (ELDER et al., 1977) suggests the occurrence of genetic exchange between virus and cell. Again, however, it is not clear that the exchange involves chromosomal, rather than episomal, virogenes.

Transfection of recipient cells by DNA from infected cells (HILL and HILLOVA, 1972) provides the strongest evidence that retroviruses are capable of integration. Very high molecular weight DNA can be infectious and the buoyant density of the infecting high molecular weight DNA is that of the bulk of genomic DNA (BAT-TULA and TEMIN, 1977) confirming the results of MARKHAM and BALUDA (1973) and additionally showing the attachment of nonviral DNA to the proviral DNA.

However, there is evidence that retroviruses do not always integrate. Unintegrated DNA can persist long after infection (VARMUS and SHANK, 1976). In one case a cell that reverted from a transformed phenotype to a nontransformed phenotype concomitantly lost proviral sequences (FRANKEL et al., 1978), implying that proviral genes can be extrachromosomal or that chromosomal genes can be selectively discarded.

Probably, it is prudent to take a conservative view, the view that proviruses integrate but that the situation is more complex than the simplest form of TEMIN's theory. Intact or nearly intact proviruses integrate because the viruses produced by transfected cells are *exactly* the ones used to produce the infected cell (HILL and HILLOVA, 1972), but it is far from clear that this is the commonest or most important mode of integration.

It is likely that recombination of free provirus with cell DNA usually results in partial integration of viral genes. The phenomenon characterizes infection of cells by DNA viruses like herpesviruses. The high incidence of abortive infections by retroviruses, even with carefully-prepared virus stocks, may result from partial integrations of provirus upon infection. Sometimes, the abortive infections are characterized by production of only selected viral proteins and occasionally the virus particles produced by the cell have only portions of the incoming viral genome. These aspects of infection of cells by RNA tumor viruses were reviewed in 1975 (GILLESPIE, SAXINGER and GALLO). Since that time, considerable effort has been spent on the phenotypic aspect of abortive infection, but few attempts have been made to define the genotype of nonproductively infected cells.

The structure of integrated proviruses has been most rigorously examined in singly-infected, cloned, virus-producing cells. In this situation a complete provirus gene representing the entire viral genome is often found. The approach which has been taken has been to treat the DNA of such virus-infected cells with selected restriction endonucleases and then to examine the size and composition of the provirus-containing DNA fragments.

The resulting picture, deduced independently by SABRAN *et al.* (1979) and HUGHES *et al.* (1978) is shown in Fig. 29. The provirus is colinear with the viral genomic RNA (see Fig. 3) in terms of the order of the *gag, pol, env* and *src* genes and carries no large sequences not found in the RNA. One unusual feature of the proviral gene is its terminal duplication. The 5′ end RNA sequence is found at both ends of the provirus as is the 3′ end RNA sequence, resulting in a long tandem repeat (LTR) unit some 300 nucleotides long. The short repeat formed at both ends of the RNA (STR) lies within LTR. In mouse sarcoma virus provirus LTR is about 600 nucleotides long (Lowy and Scolnick, personal communication) while in mouse mammary tumor virus provirus it is 1300 nucleotides long (Hager, personal communication). Interestingly, repeated mouse DNA sequences cut with the restriction endonucleases Taq 1, Bam H1 and Eco R1 are 340, 600 and 1300

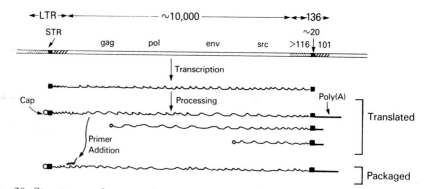

Fig. 29. Structure and proposed expression of rous sarcoma virus provirus. See text for discussion

base pairs, respectively (HOUSMAN and GILLESPIE, unpublished). Two forms of closed circular proviral DNA can be found in virus-infected cells; the closed form of the linear proviral DNA containing the terminal repeats and a smaller form lacking them (RINGOLD, SHANK and YAMAMOTO, 1978; YOSHIMURA and WEIN-BERG, 1979).

A more complicated question to answer concerns the general nature of integration events during natural infection. Do complete or partial integrations predominate? More sophisticated questions concerning the nature of integrations in special situations such as transformation or virus production come later once the first question is answered. Therefore, the cells examined should be uncloned and as little selection pressure should be placed upon them as possible. Moreover, the cells or animals used to raise the virus should approximate as closely as possible the natural host. Finally, there should be few, if any, sequences in the natural host related to the provirus.

One of the few systems that meets these criteria is the simian sarcoma virus and associated virus complex, which we shall call simian sarcoma virus. The virus was initially an endogenous virus of mice (LIEBER et al., 1975) but was transmitted to primates naturally and appears to be commonly infecting primates (WONG-STAAL, GILLESPIE and GALLO, 1975). There are few sequences in DNA of normal primates related to this virus

The natural history of simian sarcoma virus is shown in Fig. 30. The virus was originally obtained by WOLFE et al. (1971) from a cell-free extract of a fibrosarcoma

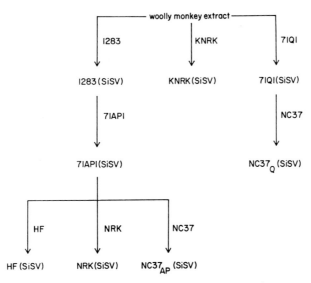

Fig. 30. Natural history of simian sarcoma virus.
See text for explanation. The nomenclature follows GILLESPIE, SAXINGER and GALLO (1975), where the designation for the cell line is followed by parentheses containing the designation for the virus. *1283* marmoset fetal lung cells, *71AP1* marmoset tumor cells, *HF* marmoset fibroblast cells, *NRK* rat kidney cells, *NC37* human lymphoid cells, *KNRK* NRK cells nonproductively transformed by Kirsten sarcoma virus, *71Q1* marmoset tumor cells

of a woolly monkey. Virus stocks commonly used have been obtained in one of three alternative ways (F. DEINHARDT, personal communication). In one instance, cultured 1283 fetal marmoset lung cells were productively infected with this extract. The resulting type-C virus-producing cells were inoculated into a marmoset which subsequently developed a tumor. The virus produced by the cultured tumor cells, the 71AP1 strain of simian sarcoma virus, was used to infect a variety of cells, including marmoset HF cells, human NC37 cells and rat NRK cells. In the second lineage, the woolly monkey extract was placed directly on NRK cells previously transformed with Kirsten murine sarcoma virus (KNRK cells). In the third lineage, the woolly monkey extract was inoculated into a marmoset #71Q1, subsequently producing a tumor in that animal. The resulting tumor cells were explanted and grown in tissue culture. The virus liberated from these cells was used to infect human NC37 cells.

Details of the hybridization results using RNA from the 71AP1 strain of simian sarcoma virus and DNA from a variety of the infected cell lines is only interpretable in the context of partial integrations being a common event in the infection of cells by this virus. This unpublished work was done in collaboration with S. Kaufmann, R. G. Smith, F. Wong-Staal and R. C. Gallo. First, the kinetics of hybridization of the viral RNA to DNA from the infected cells follows a complicated time course (Figs. 31 A and 31 B), too complicated to be accounted

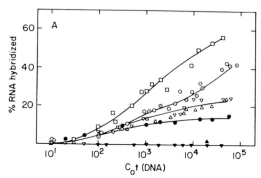

Fig. 31A. Kinetics of hybridization of RNA from simian sarcoma virus to DNA. Reactions containing 0.005 ng of ^{125}I-labeled viral RNA and 50 µg of cell DNA in 10 µl of 0.4 M phosphate buffer were incubated at 60° and withdrawn at various times. Hybridization was assayed by resistance to ribonuclease A.
□ DNA from 71AP 1 cells producing simian sarcoma virus, ○ DNA from NRK cells producing simian sarcoma virus, ⊙ DNA from KNRK cells producing simian sarcoma virus, △ DNA from NC 37 cells producing simian sarcoma virus, ▽ DNA from HF cells producing simian sarcoma virus, ● DNA from a rat spleen, ▼ DNA from NC 37 cells, ▲ DNA from marmoset cells. See legend to Fig. 30 for cell line designations

for by the presence of complete proviruses in each cell. If there were one or more complete proviruses and no incomplete ones, then the kinetics of hybridization of the viral RNA to an excess of cell DNA would follow simple second-order kinetics, going to completion over two logs of C_0t on a Britten and Kohne representation (Fig. 31 A) and being a single straight line on a Wetmur-Davidson (1968) plot (Fig. 31 B). The simplest interpretation of the kinetic data is that the RNA is

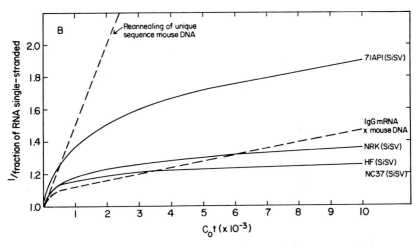

Fig. 31 B. Wetmur-Davidson representation of data of Fig. 31 A.
Some of the results from Figure 31 A have been replotted, using the coordinates of
WETMUR and DAVIDSON (1968). Hybridization of ^{125}I-labeled mouse IgG mRNA
(50% pure) was used as a single copy reference. This reference standard hybridizes
to DNA considerably slower than DNA reanneals, in this ionic environment

hybridizing to several sequences in the infected cell and the different sequences
have different repetition frequencies.

There is little hybridization to DNA from uninfected cells, save uninfected rat
cells.

The details of the stoichiometry of the hybridization reactions support the
conclusions derived from the kinetic analysis. Fig. 32 shows the results of DNA
titration experiments, where the amount of RNA is held constant and the amount
of DNA is varied, all reactions being carried out to the same C_0t value. In most
cases, relatively low amounts of DNA resulted in significant hybridization of the
RNA, but as the amount of DNA was further increased more RNA could be
hybridized. The results are presented as a double reciprocal plot because in this
formulation if all proviral sequences were reiterated the same number of times,
then a single straight line would result. In practice bi- or multiphasic curves
resulted; a fraction of the RNA could be hybridized with small amounts of DNA
(toward the right of Fig. 32) but inordinate amounts were required to approach
complete hybridization of the RNA (at the point where 1/DNA = 0, at infinite
DNA input). For example, infinite inputs of DNA might have hybridized 50 to
100% of the viral RNA but the curves are not straight lines. Thus, different parts
of the virus genome were deposited in infected cells with different frequencies. The
results with DNA from 71AP1 cells is not interpretable because of the massive
hybridization to moderately repeated DNA.

The most frequent proviral sequences differed in different lines of infected
cells, as shown by additive DNA titration experiments (Fig. 33). If the distribution
of proviral sequences were the same in all infected cell lines, then a DNA titration
experiment with two types of DNA added together would give the same curve as
the one that showed the largest hybridization by itself. If in two different cell
lines different proviral sequences predominated, then more hybridization would

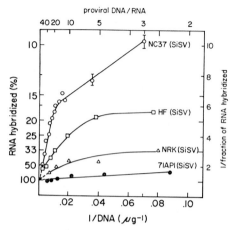

Fig. 32. Stoichiometry of hybridization of RNA from simian sarcoma virus to DNA. Hybridization mixtures containing varying amounts of cell DNA and 0.01 ng of ^{125}I-labeled viral RNA from simian sarcoma virus produced by 71AP1 cells were incubated for 160 hours (DNA concentration = 5 mg/ml, $C_0t = 2 \cdot 10^4$) in 0.4 M phosphate buffer, pH 6.8 at 60° and were then exposed to 20 µg/ml of ribonuclease A in 0.6 M NaCl for 60 min at 37°. The ratio of proviral DNA to viral RNA was calculated based on a complexity of $3 \cdot 10^6$ daltons for viral RNA and $1.5 \cdot 10^{12}$ daltons for one strand of the mammalian genomes. The results are presented as a double-reciprocal analysis. The designations for the cell lines can be found in the legend to Fig. 30

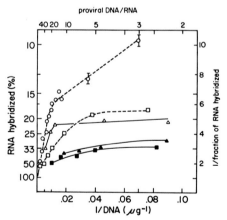

Fig. 33. Stoichiometry of hybridization of viral RNA to mixtures of cell DNA. Hybridization mixtures equal parts of DNA from two different cell lines were prepared, incubated and processed as described in the legend to Fig. 32. ▲ DNA from NC37 cells producing simian sarcoma virus and DNA from normal NIH Swiss mice, ■ DNA from NC37 cells producing simian sarcoma virus and DNA from HF cells producing simian sarcoma virus, △ DNA from normal NIH Swiss mice, ○ DNA from NC37 cells producing simian sarcoma virus, □ DNA from HF cells producing simian sarcoma virus.
The designations for the cell lines can be found in the legend to Fig. 30

be obtained by a combination of both DNAs than with either DNA alone, at least at low DNA inputs. Fig. 33 shows that in every case hybridization values obtained with small amounts of DNA were higher with a mixture of two types of DNA than with either DNA alone. For example, 20 μg of DNA from human NC37 or marmoset HF cells infected with simian sarcoma virus hybridized 12 and 17% of the viral RNA (1/DNA = .05), while 10 μg of DNA from each source hybridized 33% of the RNA. Extrapolating the DNA titration curves to infinite DNA input, considering only the contribution of the most frequently repeated proviral sequences, it is apparent that 20% of the virus RNA hybridizes to frequent sequences in either infected cell line, while 40% of the RNA hybridizes to a mixture of DNA containing frequent sequences from both sources. Therefore, the addition hybridization results indicate that different proviral sequences predominate in the two cell lines.

It seems inescapable that some viral sequences predominate in infected cells while others do not and that in different cells different sequences are most frequent. In each case, when the 71AP1 strain of simian sarcoma virus was transmitted to a new cell line genetic change took place, for the thermal stability of hybrids formed with RNA from the 71AP1 strain of virus was higher with 71AP1 parental

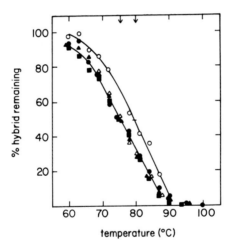

temperature (°C)

Fig. 34. Thermal stability of simian sarcoma virus RNA-cell DNA hybrids. Hybrids formed between ^{125}I-labeled viral RNA from virus produced by 71AP1 cells and cell DNA were exposed to the indicated temperature for 5 min, then ribonuclease-resistant hybrid was assayed. o DNA from 71AP1 cells producing simian sarcoma virus, ■ DNA from HF cells producing simian sarcoma virus, ▲ DNA from KNRK cells producing simian sarcoma virus, ● DNA from NC37 cells producing simian sarcoma virus, △ DNA from NRK cells producing simian sarcoma virus

DNA than with DNA from the secondarily-infected cells (Fig. 34). In every case the thermal transition was broad, indicating the presence of some proviral sequences closely related to the RNA and others more genetically distant.

The simplest way to view these results is that genetic change takes place during infection, therefore more cycles of infection produce more genetic change. Transmission is usually effected by infecting a culture at low multiplicity and allowing infected cells to accumulate by continued infection of the constantly-declining uninfected cell population. One would expect that, since proviral sequences in the secondarily-infected cells differ from proviruses in the parental 71AP1 cells, the viruses produced by the secondary cells should be different from the viruses produced by the parental cells. This can be tested in a rather sensitive fashion by RNA competitive hybridization (GILLESPIE, GILLESPIE and WONG-STAAL, 1975). Labeled viral RNA and cell DNA (fixed amounts) are mixed in the presence of different amounts of unlabeled RNA from a different virus. If the two viruses are related, then the unlabeled viral RNA depresses the amount of labeled viral RNA entering hybrids by competing for the same DNA sites. The slope of the curve obtained when the log of the fraction of the labeled RNA hybridized is plotted against the log of the amount of RNA present is a measure of the degree of relatedness between the two viruses; if they are identical, the slope is that predicted from simple isotope-dilution. By this test the genome of the virus from human NC37 cells is indistinguishable from the 71AP1 strain of simian sarcoma virus *with respect to the sequences they share* (R. G. Smith, personal communication). The only difference between the two viruses detectable by this method is that the NC37 strain lacks 20% of the sequences found in the 71AP1 strain. Fingerprints of the RNA by two-dimensional gel separation of RNase T1 digests is in support of this conclusion (W. PRENSKY, personal communication).

Direct hybridization of labeled RNA from the NC37 strain of virus also supports the notion that the genomes of the two viruses are identical except for the extra sequences in the 71AP1 strain. Like RNA from the 71AP1 strain, RNA

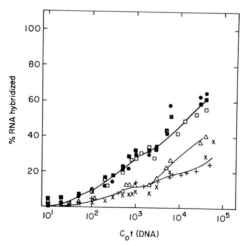

Fig. 35. Comparison of hybridization kinetics using RNA from simian sarcoma virus produced by 71AP1 and NC37 cells.
Procedure are outlined in the legend to Fig. 31A. □ and ■ RNA from 71AP1 virus hybridized to DNA from 71AP1 cells producing simian sarcoma virus; ×, + RNA from the 71AP1 virus hybridized to DNA from NC37$_{AP}$ and NC37$_Q$ cells producing simian sarcoma virus, the AP and Q cell strains, respectively (see Fig. 30); ● RNA from the NC37 virus hybridized to DNA from 71AP1 cells producing simian sarcoma virus; △ RNA from the NC37 virus hybridized to DNA from NC37 cells producing simian sarcoma virus

from the NC37 strain hybridizes best and fastest to DNA from 71AP1 cells and actually hybridizes more poorly to DNA from infected NC37 cells, the cells producing it (compare Figs. 35 and 30). The hybridization kinetics curves using RNA from the two viruses are superimposable, except that RNA from the NC37 strain hybridized slightly better to DNA from its own host at high C_0t values. Similarly, the thermal stabilities of hybrids formed with RNA from the NC37 strain and provirus DNA were like those formed with RNA from the 71AP1 strain. The hybrids with the highest thermal stability were those with DNA from 71AP1 cells. Their t_m was at least 4° higher than the t_m of hybrids formed with DNA from other infected cells. Thus, the genetic variation among proviruses from different cells infected with simian sarcoma virus, is contradicted by the virtual identity of the viruses produced by the different cells.

Let us again try to take the conservative view. It can be postulated that simian sarcoma virus is in fact two viruses, a true sarcoma virus capable of transforming fibroblasts and causing tumors in animals and, in addition, a helper virus required to supply replication functions lacked by the (defective) sarcoma virus. This is, after all, the conventional view of mammalian sarcoma viruses. Certainly, helper virus, or more accurately replication-competent, transformation-defective virus has been purified from the sarcoma virus stock (WOLFE et al., 1972). Moreover, the 71AP1 strain is a transforming virus and the NC37 strain is not (S. Mayassie, K. Harewood and Z. Sallahudin, personal communication). The extra sequences in the genome of the 71AP1 virus could simply be sarcoma-specific sequences, sequences that are missing in the nontransforming NC37 strain.

Unfortunately, the conservative view is insufficient this time. The hybridization results are not explained by sorting out two viruses. The complexity of the provirus is indicative of many different but related forms of provirus with different parts in different frequencies. Moreover, if two (or more) viruses were being sorted, the nature of the provirus would dictate the nature of the genome of the virus produced and this is not the case. Finally, in the specific case of sarcoma and helper viruses, the helper predominates 1000-fold over the sarcoma virus, at least as measured biologically (WOLFE et al., 1972).

No, we are pressed to an unconventional view. We must explain that pieces of an incoming virus genome can be integrated at different frequencies, that genetic change can take place in the process and that, despite the fracturing, multiplication and change of the proviral sequences, the virus can come out only minimally changed—a deletion, after all, is ordinarily only a single genetic event.

The simplest explanation is that partial integration is the rule, rather than the exception. Repeated integrations can probably occur once a cell has become infected. It follows from the data that only certain of the integrations are expressed by production of infectious virus; probably the "complete" integrations and those complemented by host functions, as in the case of infections that produce recombinant viruses. That is not to say that the partial integrations are without biological consequence. Indeed, they may play a major role leading to transformation.

It is probably worth considering, for example, that a provirus integrated within or near a repeated sequence might be readily translocated to alternate chromosomal sites. Should the translocated element intervene within a unit of transcription, aberrant expression of that unit might result. The protovirus theory

(TEMIN, 1970) requires such a translocation mechanism to create a tumorigenic virus from an endogenous virogene transcript. The involvement of repeated DNA not only provides a logical mechanism for the translocation process but also provides a satisfactory explanation for the expression of repeated DNA in tumors (SHEARER and SMUCKLER, 1971 and 1972; SMUCKLER, 1973).

One piece of preliminary evidence bearing on this point is shown in Fig. 36. For this experiment human DNA (probably DNA from any primate will suffice) was digested with a restriction enzyme that cuts repeated DNA at regular intervals, leaving a large number of DNA fragments 340 base pairs in length (DONEHOWER and GILLESPIE, 1979). The digested DNA was fractionated by electro-

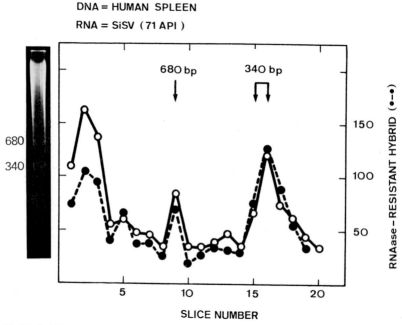

Fig. 36. Hybridization of RNA from simian sarcoma virus to human DNA fragments produced by Endo · Xba I.

Ten micrograms of DNA from a normal human spleen was digested with the restriction endonuclease Xba I. The DNA was fractionated according to size by electrophoresis through gels of 2% agarose. DNA was stained with ethidium bromide and visualized under ultraviolet light. The inset to the left of the graph shows a typical gel electrophoretogram, noting the position of repeated DNA fragments 340 and 680 base pairs long. This repeated sequence represents under 1% of the human genome

The gel was sliced and the slices were dissolved in 2 vols of saturated NaI. DNA was bound to glass and collected by centrifugation (VOGELSTEIN and GILLESPIE, 1979). DNA was released from the glass in 0.4 M NaOH. The solution was neutralized with acetic acid and the denatured DNA was immobilized on nitrocellulose membranes. Each filter was incubated in 50 microliters of 50% formamide and 0.5 M NaCl containing 10^5 cpm of ^{125}I viral RNA at 37° for 20 hours. After hybridization unreacted RNA was washed away and the filters were counted in a gamma counter (open circles). The filters were then incubated in 20 µg/ml of ribonuclease A, free RNA was washed away and the hybridized RNA remaining on the filter was measured (closed circles). The scale for the open circles is four times that for the closed circles

phoresis through agarose. The gel was then sectioned into slices and the DNA was recovered from each slice and hybridized to DNA from simian sarcoma virus. The hybridization system was constructed so that only repeated DNA sequences would participate, providing they had homology to sequences in the viral RNA. As can be seen from Figure 36, about half of the hybridization response observed was to DNA about or exactly 340 base pairs in length. The result indicates that simian sarcoma virus, at least the strain tested has a genome containing a cell-derived repeated DNA sequence. This does not prove that the repeated DNA is a potential viral integration site, but it is consistent with that idea.

HUGHES *et al.* (1978) determined the restriction map of integrated provirus from twelve clones of cells infected by Rous sarcoma virus. They found three with large internal deletions. The three deletions all deleted a common site including the *env* gene and approaching the terminus of the *src* gene. They found no evidence for translocation of provirus once established in the cell. It is not known whether the deletions were formed during provirus synthesis from RNA or during integration, but the former seems most likely (SHIH *et al.*, 1978).

In any event, interaction between the DNA provirus of an infecting particle and the chromosome of the recipient cell can clearly create recombinant viruses carrying cellular genes and evidence is mounting that these recombinant viruses are more oncogenic than their cell gene-deficient viral precursors.

V. Horizontal Transmission of Retroviruses Among Animals

The common view holds that class 1, endogenous, type-C RNA viruses are transmitted as virogenes, vertically through the germline and are thus confined to interbreeding members of a group of animals (usually a species), while class 2 viruses can be transmitted horizontally among animals. Until recently, it was thought that horizontal transmission was also confined to members of the same species. There is ample evidence to show that horizontal and vertical transmission of RNA tumor viruses occurs among members of a species, but new evidence suggests that interspecies horizontal transmission also occurs.

Intraspecies horizontal transmission of RNA tumor viruses has been documented with several species, notably chickens (ROUS, 1911), mice (RASHEED, GARDNER and CHAN 1976; HARTLEY and ROWE, 1976), cats (ESSEX *et al.*, 1971; HARDY *et al.*, 1973; JARRETT *et al.*, 1973) and cows (Piper *et al.*, 1975). Chickens carrying avian myeloblastosis virus have been known to infect entire flocks. Cats carrying feline leukemia virus readily infect other cats in a household, some developing leukemia. Transmission among wild mice also occurs, though not as readily as with inbred laboratory mice. The articles cited above describe recovery of infectious virus from recipient animals. When virus infection is assayed by the production of antibodies against virus-specific proteins by recipient animals, intraspecies horizontal transmission of RNA tumor viruses appears to be even more common. Thus, normal gibbons housed in the same colony with leukemic gibbons often produce antibodies against proteins of the class 2 virus isolated from the leukemic

gibbon even though virus cannot be isolated from the recipient animals (KAWA-
KAMI et al., 1973).

The presence of proviral DNA sequences in naturally-infected animals is
poorly documented. Mice inoculated with Rauscher leukemia virus contain
proviral sequences in tumor tissue that are absent in normal tissue (SWEET et al.,
1974) and AKR mice carry "proviral" sequences not found in nonleukemic mouse
strains (CHATTOPADHYAY et al., 1974), but these are hardly natural infections. In
the case of web tumors in chickens induced by Rous sarcoma virus a large fraction
of the virus genome is found in DNA of the tumor (NEIMAN, 1972). The cases
known to us of proviral DNA spontaneous tumors of mammals include several
tissues of a gibbon that spontaneously developed acute lymphocytic leukemia
(GALLO et al., 1978), lymph nodes and blood leukocytes of leukemic cows (CALLA-
HAN et al., 1976; KETTMAN et al., 1976), sarcomas of cats (RUPRECHT, GOODMAN
and SPIEGELMAN, 1973), leukemic tissues of cats (KOSHY et al., 1979) and blood
leukocytes from leukemic humans (BAXT and SPIEGELMAN, 1972; WONG-STAAL,
GILLESPIE and GALLO, 1976). None of these cases can be safely classified as
interspecies horizontal transmission (see below).

Viruses which originated from virogenes in a particular animal species often
contain genomes that are most closely related to sequences in DNA from their
natural host, providing the virus infected only the one animal species during its
evolutionary history. Rauscher mouse leukemia virus is a case in point (Fig. 7).
As we shall see later, viruses that have crossed species can bear preferential
relatedness either to their progenitor animal, as in the case of simian sarcoma
virus, a virus that originated from mice but that now infects primates or to their
new natural host, as in the case of feline leukemia virus, a virus that originated in
rats but now infects cats. Viruses that have remained within the bounds of a
single species for a long period of time are generally more closely related to genes
in that species than to genes in other species. RNA from Rous sarcoma virus
(class 2), avian myeloblastosis virus (class 2) and RAV-0 (class 1) hybridizes to a
greater extent to an excess of DNA from normal chickens than to DNA from
other birds (KANG and TEMIN, 1974; TEREBA, SKOOG and VOGT, 1975; SHOYAB
and BALUDA, 1975; FRISBY et al., 1979). RNA from Rauscher and Moloney leukemia viruses of mice (class 2) and mouse mammary tumor virus and the AKR mouse
leukemia virus (intermediate between class 1 and class 2) hybridize better to DNA
from normal mice than to DNA from rats or other animals (GILLESPIE et al.,
1973; BENVENISTE and TODARO, 1974; DROHAN et al., 1977). RNA from an
endogenous guinea pig virus (class 1) hybridizes better to DNA from guinea pigs
than to other rodents (NAYAK and DAVIS, 1977). Finally, RNA or DNA copies
of it derived from the baboon endogenous virus hybridize better to DNA of
baboons than to DNA of other primates (Fig. 6) (BENVENISTE and TODARO, 1976;
DONEHOWER, WONG-STAAL and GILLESPIE, 1977). The results of such studies are
more convincing when larger numbers of animals are examined and a phylogenetic gradation can be shown. In this regard, the data with mouse mammary tumor virus and the guinea pig virus are the least complete.

The results of molecular hybridization using the RNA genomes of the viruses
cited above indicate that those viruses have been maintained within the confines
of their progenitor host species. Other retroviruses have not remained confined in

this way. BENVENISTE and TODARO have amassed evidence that genes related to the genomes of feline leukemia virus and the RD114 virus were introduced into cats from other animals (BENVENISTE and TODARO, 1974). SCOLNICK and his coworkers have shown that Kirsten, Moloney and Harvey murine sarcoma viruses contain a mixture of mouse- and rat-derived genomic sequences or two different mouse-derived sets of sequences (SCOLNICK et al., 1973, 1975; SCOLNICK and PARKS, 1974; see also PANG, PHILLIPS and HAPAALA, 1977; ROY-BURMAN and KLEMENT, 1975; ANDERSON and ROBBINS, 1976 and, the following section). Similarly, simian sarcoma virus and gibbon ape leukemia virus entered primates from rodents (BENVENISTE and TODARO, 1973; WONG-STAAL, GALLO and GILLE-SPIE, 1975). Bovine leukemia virus appears not to have originated in cows (CAL-LAHAN et al., 1976; KETTMAN et al., 1976). Finally, BENVENISTE and TODARO (1975) furnished evidence that pig leukemia and endogenous viruses were originally mouse viruses and were transmitted to pigs within the last few million years.

Most of the cases of interspecies transfer of RNA tumor viruses that we know today were reviewed by TODARO, BENVENISTE and SHERR in 1976. Let us briefly examine the kind of evidence that prompted them to such an unusual and important conclusion by considering the viruses that have been transmitted to cats from other animals in nature.

The virus called RD114 was isolated from a kitten which had been inoculated with extracts of a human rhabdomyosarcoma during attempts by HUEBNER, McALLISTER and colleagues to isolate oncornaviruses from extracts of human tumors (McALLISTER et al., 1972). Intense work on this virus quickly showed that it was an endogenous feline virus with a genome having complete homology with genes in normal cats (NEIMAN, 1973; GILLESPIE et al., 1973; RUPRECHT, GOODMAN and SPIEGELMAN, 1973; OKABE, GILDEN and HATANAKA, 1973; BALUDA and ROY-BURMAN, 1973). Thus, it was abandoned as a candidate human virus but it has become an important keystone in our understanding of the origin and evolution of RNA tumor viruses.

Near the same time KALTER and MELNICK isolated an endogenous virus from baboons (MELNICK et al., 1973). The two groups carried out the initial character-ization of the virus(es) with TODARO and his colleagues (BENVENISTE et al., 1974—with S. KALTER; TODARO et al., 1974—with J. MELNICK). TODARO and BENVENISTE and their colleagues quickly determined that the baboon endogenous virus carried proteins that bore no immunological relatedness to any known virus, except RD114 (see TODARO et al., 1976) and that the genomes of the two viruses were much more closely related than would have been predicted from the phylogenetic distance between cats and baboons. BENVENISTE and TODARO then discovered that there were sequences in the DNA of baboons and related pri-mates with more than expected homology to the RD114 genome (BENVENISTE and TODARO, 1974) and there were genes in cats unusually related to the genome of the baboon endogenous virus (BENVENISTE and TODARO, 1974). Other animals, e.g. rodents, edentates and new world monkeys may have related genes but they were so distant from the viral genomes that no hybridization was found.

This set of data led BENVENISTE and TODARO to propose that a virus of baboons had been transmitted to cats or that a virus of cats had been transferred to baboons. They then noticed that the genes related to baboon endogenous virus showed a phylogenetic gradation among primates that could be traced back some 30 million years (BENVENISTE and TODARO, 1976; see also DONEHOWER, WONG-STAAL and GILLESPIE, 1977) while the RD114 genes of domestic cats could only be traced back some 3—5 million years (BENVENISTE and TODARO, 1975). That is, the RD114 genes are present in a small number of closely related cats, cats which (presumably) were derived from a common ancestor 3—5 million years in the past but they are absent, *totally absent*, in all other cats. Thus, BENVENISTE and TODARO concluded that the transfer went from baboons to cats about 3—5 million years in the past.

The transfer of genes among species, using viruses as vectors, must occur in stages. To facilitate future discussions let us divide the process of transspecies infection as follows (Fig. 37). A virus produced by one species infects an individual

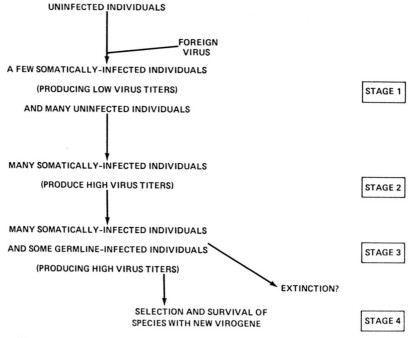

Fig. 37. Postulated stages in interspecies transmission of retroviral genes

of another species so that the infected individual produces a virus capable of infecting the second species with a reasonable frequency. A few individuals of the second species become infected and produce low titers of virus. A second stage develops as the virus becomes better and better adapted to its new host, increasing the number of individuals infected and the titer of viruses produced by the infected individuals. Eventually, the virus titer becomes sufficiently high that the probability of a germline infection is appreciable. The occurrence of one or more germline infections marks the onset of stage 3. Such an infection of germline tissue, leading to virogenes segregating in a Mendelian fashion has been accomplished by JAENISCH (1976, 1977).

But these events are only the beginning, for during a fourth stage the individual infected in its germline would be selected for with the emergence of a new "species" of animal. Especially in the case of the RD114 transfer this seems so, for every tissue of every domestic cat examined carries RD114 virogenes. The RD114-positive cats form a phylogenetically-defined group, as do the RD114-negative cats. To date, there are no spurious results, *e.g.* cats in the RD114-negative group with RD114 sequences or *vice versa*.

It is difficult to take a conservative view. Were animals with new viral sequences in the germline not actively selected for in a Darwinian sense, then at best *some* members would contain the viral genes and they would be heterozygous. The copy number of RD114 genes in different members of the species would vary and also, some would be zero. In practice, cats with the RD114 genes appear homozygous and all tissues of all cats so far examined have about 10 copies of the RD114 virogenes (BENVENISTE and TODARO, 1975).

It is not sufficient to propose that animals carrying the RD114 virogenes are actively selected for, because RD114-positive cats not only carry RD114 virogenes; they are homozygous. RD114 negative cats are also homozygous in lacking the viral genes. Fig. 38 presents three possible routes to homozygosity in the RD114 system. The three schemes include variations that seem to fit individual cases of transfers detected so far. Generally, a germline infection creates a hetero-

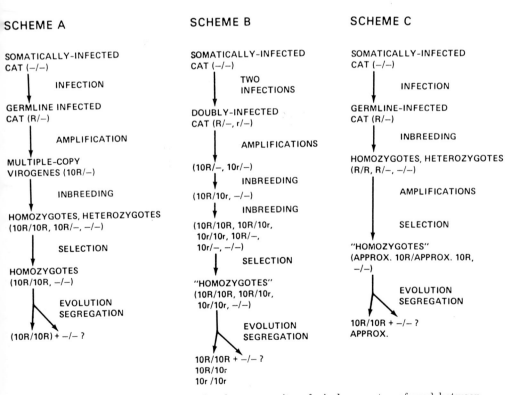

Fig. 38. Mechanisms for generating homozygosity of viral genes transferred between species

zygote. The viral genes are amplified to a linked (but not necessarily immediately adjacent) series of multiple-copy virogenes and homozygotes are added to the population by inbreeding. Through evolution and segregation pure lines are established. Pseudohomozygosity can be generated by creating two or more germline-infected individuals independently (scheme B). If related viruses are involved this can lead to mixed genotypes that appear homozygous by molecular hybridization (and possibly functionally). Alternatively, homozygosity can precede amplification (scheme C). Mechanisms of unequal crossover can change copy numbers in homozygotes, the final copy number being selected at a functional level (RITOSSA, 1968).

Interestingly, there is one feature to all three schemes that is common. The heterozygote is selected against. It did not survive. RD114-positive and RD114-negative cats interbreed, producing fertile offspring in captivity, but naturally no heterozygotes exist. Possibly, two functional and different virogenes produce contradictory signals during development and are thus incompatible. In any event, it will be interesting to see whether the principle of homozygosity is general or restricted to cats.

RD114 is not the only virus to have entered cats from another animal. Feline leukemia virus was originally an endogenous rat virus and was transmitted to cats at about the same time that RD114 entered cats from baboons (BENVENISTE, SHERR and TODARO, 1975). The same species of cats that carry RD114 virogenes also carry sequences related to the feline leukemia virus and the remaining cats lack both types of viral sequences. It is somewhat staggering to think that genes can be transferred from one animal to another and be selected for, but that it should happen twice, essentially simultaneously, seems incredible.

The parallel between RD114 and feline leukemia virus should not be extended too far. RD114 is now an endogenous cat virus; feline leukemia virus clearly is not. The RNA of feline leukemia virus contains some sequences not found in DNA of normal cats. It is not difficult to stipulate that a gene transferred from one animal to another might be expressed in the new animal as RNA once the fact of transmission itself is established. It is more of a problem to understand how, after transfer, there could exist a new DNA sequence and new RNA similar in base sequence but *nonidentical* to the new DNA.

Fig. 39 attempts to come to grips with this problem. It shows the four stages of transspecies infection with specific reference to the rat-to-cat transfer which led to the production of feline leukemia virus. In the first stage an ancestor to the present-day rat produced retroviruses. One of these viruses infected a cat in such a way that the infected cat cell produced infectious viruses. Probably cats and other animals were and are commonly infected by retroviruses from other species but only those infections that are productive are involved in interspecies transfer of viral genes into germlines.

During the second stage of the transspecies infection leading to feline leukemia virus the newly-transferred virus adapted to efficient growth in its new host. Members of the virus population repeatedly interacted with the cat genome during infection and they became genetically modified by this interaction and by other mutational events. In this way a population of related, nonidentical viruses could have arisen in the cat, much in the way that today in primates the simian

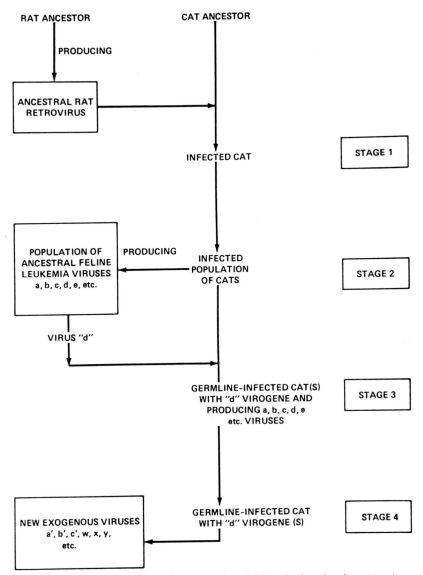

Fig. 39. Postulated stages in the transfer of feline leukemia virus to cats

sarcoma virus and its associated virus is a population of related, nonidentical viruses.

At this second stage the only DNA sequences in cats related to feline leukemia virus is found in infected tissue. If cats were infected by a population of feline leukemia viruses and if the RNA probes used to examine cat DNA for viral related sequences came from one or a few of these viruses, then what might be most frequently detected would be DNA related to the RNA probe but not identical to it. This is in fact what is found. However, to explain the observation that all cats examined have the viral-related DNA one would have to postulate that infection is much more widespread than is presently thought (virtually all tissues of all cats would be infected) and that if enough cats were examined some would be found that lacked DNA related to the feline leukemia virus. Indeed, KOSHY *et al.* (1979) examined over 100 cat tissues and discovered some with unusually low values. It will be interesting to examine the viral-related sequences in these tissues in detail, with respect to number of copies and which portion of the viral genome is hybridizing to DNA of different cats, for during stage 2 of transspecies infection there should be a randomness not seen at later stages. However, there are so many logical problems with assuming that feline leukemia virus is in stage 2 — the feline leukemia virus negativity of RD114-negative cats, the presence of viral-related sequences in specific-pathogen-free cats, for examples — that for this monograph we consider the stage 2 model unlikely.

During stage 3 one or more of the members of the ancestral population of feline leukemia viruses infected germline cells of a cat(s) and the descendants of this cat increased in frequency in the cat population. At this stage both the virus population and the cat population were complex in composition. The population of ancestral viruses, somatically propagated, would persist until their replication is controlled by the cat. At this stage some members of the cat population carry the viral genes as a Mendelian trait, others do not. Some members are infected somatically, others may not be. Using an RNA probe from a cloned feline leukemia virus only related DNA sequences would be seen in the cat and this is the case.

However, a stage 3 model is not consistent with all of the present results, for it predicts that the majority of feline leukemia virus-positive cats will be heterozygous and that there will still exist cats that are homozygous-negative for feline leukemia virus genes. The limited data available suggests that the feline leukemia virus-related DNA in cats is homozygous (BENVENISTE, SHERR and TODARO, 1975).

During stage 4 germline-infected cats (homozygous) flourished and reached a state where they comprised a phylogenetically defined group of cats. Thus, all cats within the group carried the new viral gene and essentially all cats outside the group lacked them. In addition viruses replicating somatically could introduce new DNA into cats with a frequency that is unpredictable. The virus population might be simple, if the new virogene is expressed and if the species is rid of the ancestral virus population as in the case of RD114, for example. Or it can remain complex if the ancestral virus population persists. Moreover, if the new virogene is expressed there is the potential to package the RNA transcripts and create new viruses from them through genetic change as TEMIN (1970) proposed. Again the DNA of normal cats contains sequences related but not identical to the RNA probe of a cloned feline leukemia virus until the virogene transcript is discovered and utilized. The copy number of the viral sequences will appear more uniform among cats than in the cases of earlier stages of transspecies transmission. And finally, after segregation and selection homozygous cats can make up the bulk or all of the cat population.

We are still left with the perplexing problem of why all RD114-positive cats are positive for DNA related to feline leukemia virus and *vice versa*. Probably there was a causal relationship. BENVENISTE and TODARO (1975) supposed that the introduction of RD114 into cats facilitated infection by an ancestor of feline leukemia virus. Of many possible mechanisms for generating dually homozygous multiple-copy sets of new genes in a species two of the more likely schemes are shown in Fig. 40. A germline infection of a cat by RD114 is closely followed by an infection with feline leukemia virus (scheme D). In the simplest version the feline leukemia virus integrates near the RD114 gene. The two linked genes can then be amplified to an alternating unit and homozygosity can be achieved by

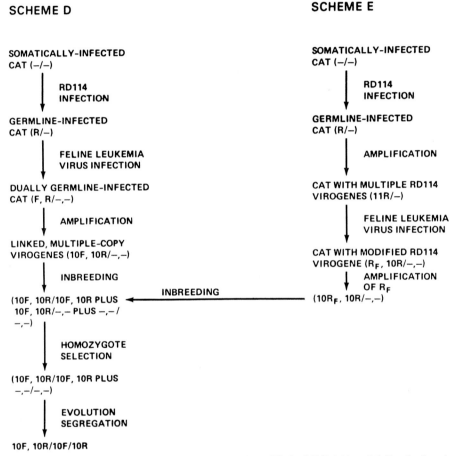

Fig. 40. Mechanisms for generating homozygosity of linked RD114 and feline leukemia virus genes in cats

inbreeding. It is more complicated to achieve distinct, homozygous groups of cats either positive for both virogenes or negative for both virogenes if the genes enter different cat chromosomes.

The scheme D model implies that the RD114 integration made the site of integration unusually susceptible to subsequent integration in the same vicinity by feline leukemia virus. This could be a direct potentiation as indicated in scheme E of Fig. 39 or it could be an indirect mechanism. By a direct potentiation we mean that an RD114 virogene was the integration site for feline leukemia virus.

In the simplest case, RD114 would have infected the germline of a cat and the new genes would have been amplified. Cats heterozygous for the multiple-copy virogene would exist and feline leukemia virus would integrate into one member of the virogene family. This virogene member would then be amplified and made homozygous by inbreeding.

It should be emphasized that if feline leukemia virus inserted within RD114 virogenes (Fig. 41, Model 2, virogene R_f) the hybrid virogene must be expressed only rarely, if at all. As far as is known, feline leukemia virus RNA has no RD114 sequences in it. It has to be stipulated in this model that infectious feline leukemia virus is propagated somatically, not by expression of Mendelian cat genes. If feline leukemia virus integrated elsewhere *in toto* (Model 1) or fragmented within a virogene (Model 3) or elsewhere, it might be expressed but not as any of the commonly-used laboratory strains of feline leukemia virus. Note in Fig. 41

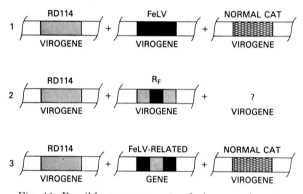

Fig. 41. Possible arrangements of virogenes in cats

that a normal cat virogene is postulated, apart from RD114 virogenes and virogenes related to feline leukemia virus. If such virogenes exist in cats, as they do in other animals, then they will be found in *all* cats, not only in the RD114-positive, feline leukemia virus-positive group.

Alternative to an integration of feline leukemia virus within RD114 genes, the infection of cats by RD114 could have had an *indirect* effect on facilitating integration of feline leukemia virus. L. Donehower (unpublished) discovered a newly evolved repeated sequence in DNA of cats, a sequence confined to cats positive for RD114 virogenes. Fig. 42 shows that RNA extracted from feline leukemia virus carries sequences that hybridize to this 510 base pair repeated DNA. Possibly, infection of cats by RD114 caused amplification of repeated DNA which served as an integration site for the feline leukemia virus. This particular isolate of feline leukemia virus was grown in cat cells, so it is not certain that the sequences in the RNA which hybridized to the repeated DNA were part of the viral genome as opposed to a cell RNA contaminant. Further studies will clarify this point and the general idea that infection by RD114 caused a local chromosomal change potentiating integration of feline leukemia virus.

To summarize this section, it seems inescapable that interspecies transmission occurred naturally and that the consequences of such a transfer can include the establishment of a new gene in the species and cancer. It is important to try to understand the basis of the gene transfer process and the aspects that contribute to evolution, on the one hand, and cancer on the other. Does the mobility of retroviral genes among species indicate the possibility of mobility of the same genes within a given genome (TEMIN, 1971)? If so, do repeated DNA sequences contribute to the mobility? Does the expression of repeated DNA and the cyto-

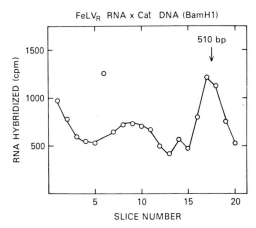

Fig. 42. Hybridization of RNA from feline leukemia virus to repeated cat DNA. Ten micrograms of DNA from a domestic cat was digested with the restriction endo-nuclease, EndoR · Bam H 1. The DNA was fractionated according to size by electro-phoresis through gels of 2% agarose. The gel was sliced and DNA removed and immo-bilized to nitrocellulose membranes as in Fig. 36. Each filter was incubated with hybridization solution containing 10^5 cpm of ^{125}I viral RNA. Ribonuclease-resistant hybrids were assayed

plasmic appearance of high molecular weight RNA in tumors suggest pathological aspects of RNA synthesis in neoplasia? Has the transfer of retroviruses to humans occurred and if so, is cancer a possible consequence? Hopefully, research in the next few years will provide answers to some of these questions.

VI. Relatedness Among Retroviruses: Recombinant Viruses

The general statement has been made that RNA tumor viruses from different species are related to one another and are interrelated in the same way their natural hosts are related (BENVENISTE and TODARO, 1973; MILLER et al., 1974; OKABE, GILDEN and HATANAKA, 1973; HAAPALA and FISCHINGER, 1973; EAST et al., 1975). The validity of this statement might be expected, considering the evidence that type-C RNA viruses originated from cell genes. On the other hand, several facts complicate the homology pattern considerably, including the produc-tion of several, unrelated viruses by one animal, transmission of viruses between animal species and recombination among viruses.

Interviral homologies have been assessed by two molecular hybridization techniques; by hybridizing partial DNA transcripts of RNA from one virus to genomic RNA of several others and by hybridizing viral RNA to cell DNA and competing the hybridization with an excess of unlabeled RNA from other viruses. The DNA transcript (cDNA) technique itself has two variations. In the first, small amounts of labeled cDNA are hybridized to vast quantities of viral RNA and the fraction of the cDNA hybridized is measured. In the second, a small amount of labeled viral RNA is hybridized to a vast amount of cDNA and the percent of the viral RNA hybridized is assayed. All three methods have yielded

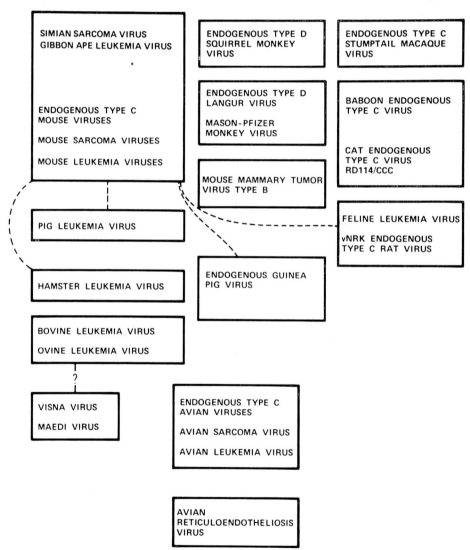

Fig. 43. Diagram of relatedness among retroviruses.
See text for details. Viruses enclosed in a box are demonstrably related to one another
by molecular hybridization and comparative serology

equivalent results, despite considerable hullabaloo over whose method was superior. The actual numbers of percentages of RNA or cDNA hybridized or of RNA competed varied with experimental conditions. Most studies included many viruses, however, and it is the pattern of homology that gives informations. These patterns are presented in Figs. 43—46. They are simplified, but as far as we know, they are accurate.

Fig. 43 gives an overview of the relatedness among avian and mammalian retroviruses, stressing type-C viruses and the leukemia and sarcoma viruses derived from them, but in some cases including type-A (mouse), type-B (mouse) and type-D (primate) viruses as well. Each species or group of closely related animals (e.g. primates) can give rise to many unrelated endogenous viruses and can harbor different exogenous, class 2 viruses (GILLESPIE, SAXINGER and GALLO, 1975). In Fig. 43 a box encloses related viruses obtained from the same animal species. These viruses are unrelated to viruses in other boxes (less than 2% genomic homology) unless homology is indicated by a connecting bar. Most of the data is from BENVENISTE and TODARO, 1973; MILLER et al., 1974 and EAST et al., 1975. In the mouse and rat category, distinct viruses have recombined to form viruses related to both progenitors (see SCOLNICK references and later). The animal groups that have been studied most intensely are the birds, mice and rats, cats, and primates. These groups are considered individually.

RAV$_0$, ILV-UV, I-ILV
RAV-2, RAV-60, RAV-61, Myeloblastosis-associated virus
Avian osteopetrosis virus
Avian myeloblastosis virus
Rous sarcoma virus
B 77 Avian sarcoma virus

Chicken syncytial virus
Reticuloendotheliosis virus
Trager duck necrosis virus
Duck infectious anemia virus

Fig. 44. Diagram of relatedness among avian retroviruses

Virtually all type-C viruses obtained from chickens are closely related to one another (Fig. 44) (KANG and TEMIN, 1973; NEIMAN et al., 1974; HAYWARD and HANAFUSA, 1975; SHOYAB and BALUDA, 1976). KANG and TEMIN (1973) studied three endogenous viruses—subgroup E—and found them indistinguishable. RAV-0 is spontaneously produced by line 100 chicken cells (VOGT and FRIIS, 1971) and is a class 1 virus (NEIMAN, 1973). ILV-UV and I-ILV were induced from normal chicken cells by WEISS et al. (1971). KANG and TEMIN were unable to distinguish these viruses from the weakly sarcomagenic RAV-61 virus, produced by passing the Bryan high titer strain of Rous sarcoma virus through pheasant cells (HANAFUSA and HANAFUSA, 1973) or from the strongly sarcomagenic viruses, the Schmidt-Rupin strain of Rous sarcoma virus and the B77 avian sarcoma virus. Other studies routinely describe sequences in the sarcomagenic viruses that are absent in the "endogenous" helper viruses (NEIMAN et al., 1974; HAYWARD and HANAFUSA, 1975), the so-called src sequences (STEHELIN et al., 1976). RAV-2 is a class 2 avian virus (HAYWARD and HANAFUSA, 1975) which, when passed through normal chicken cells "rescues" a new virus, RAV-60 (HANAFUSA, HANAFUSA and MIYAMOTO, 1970). RAV-2 is related but nonidentical to the class 1 virus, RAV-0 and RAV-60 seems to be genetically intermediate (HAYWARD and HANAFUSA, 1975).

In this same group of viruses belong the leukemogenic chicken virus, avian myeloblastosis virus (SHOYAB and BALUDA, 1975) and an avian osteopetrosis virus obtained from a stock of avian myeloblastosis virus (SMITH and MUSCOVICI, 1969; SMITH, DAVIDS and NEIMAN, 1976). A field isolate of avian lymphomatosis virus is also closely related (SMITH, DAVIDS and NEIMAN, 1976). The three viruses named above are all class 2 viruses, containing some sequences not found in the DNA of uninfected chickens but, like the avian sarcoma viruses they contain genomes that are largely related to the genes of the true endogenous chicken virus, RAV-0 (NEIMAN, 1972; NEIMAN, PURCHASE and OKAZAKI, 1975; SHOYAB and BALUDA, 1976).

It seems routine that avian viruses represent a mixture of genetically related nonidentical individuals, until cloned, like the situation described earlier for simian sarcoma virus (pp. 51—59). Even after cloning, new related viruses can accumulate (Bishop, personal communication). It is probably inaccurate to think of RNA tumor viruses as stable genetic entities, especially in their setting, in their natural host, where recombination with cell genes occurs frequently (see later).

There exists a second, separate group of avian type-C viruses, called the reticuloendotheliosis virus group. As far as we can tell, they are grouped because of their homology (KANG and TEMIN, 1973) rather than because they have a common origin or common pathology. The members of the reticuloendotheliosis virus group include chicken syncytial virus, avian reticuloendotheliosis virus, Trager duck necrosis virus and duck infectious anemia virus.

The chicken syncytial virus was isolated by COOK (1969) from the CAL-1 strain of chicken with Marek's disease. Marek's disease chickens also release RAV-0 endogenous RNA virus as well as a herpes DNA virus. Chicken syncytial virus is not antigenically related to the avian leukosis viruses (COOK, 1969).

Avian reticuloendotheliosis virus was isolated from adult turkeys by TWIEHAUS and ROBINSON (1965). It was adapted to young chicks but is unrelated antigenically to avian leukosis viruses (THEILEN, ZEIGEL and TWIEHAUS, 1966). The virus is a typical type-C RNA tumor virus but the purified virus is not nearly as pathogenic as extracts of spleens from infected birds (BAXTER-GABBARD et al., 1971).

Trager duck necrosis virus and duck infectious anemia virus were both discovered as contaminants of malarial parasite, Plasmodium lophurae (TRAGER, 1959; McGHEE and LOFTIS, 1968; LUDFORD, PURCHASE and COX, 1972). Similar viruses were found in several different malarial parasites of birds (DEARBORN, 1946). Plasmodium lophurae was originally isolated from a ring-necked pheasant (COGGESHALL, 1938).

All of these viruses are highly related, even though they were isolated from birds that are related to one another only distantly from the point of view of the homology of their DNA (KANG and TEMIN, 1973, 1974). The Trager duck necrosis virus lacks homology with DNA from chickens, ducks, pheasants or turkeys and presumably the other viruses will behave similarly. Two of the viruses are associated with a parasite that is horizontally-transmitted by insects and a third is in association with a horizontally-transmitted DNA virus. The fourth is more infectious in crude extracts than when purified, suggesting it, too, exists in association with another agent. Possibly, a common ancestor of these viruses was transmitted to domestic birds from another animal but, as KANG and TEMIN (1974) emphasize, their genetic origin is not clear.

Fig. 45 interprets the relatedness of selected chicken RNA tumor viruses to genes in normal chickens. RAV-0 is considered to be a direct RNA transcript of a chicken virogene. It may be altered somewhat since it is produced as an extracellular virus, but it is not tumorigenic. RAV-2 is pictured as a descendant of RAV-0, having evolved away from the chicken virogene by some undefined means. RAV-60 is a recombinant virus (HAYWARD and HANAFUSA, 1975) formed by passing RAV-2 through normal chicken cells (see later).

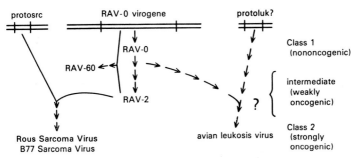

Fig. 45. Diagram of relatedness of some avian retroviruses to cell genes

Rous sarcoma virus is a product of recombination between RAV-0 or a descendant of it and a cell gene termed *"protosrc"* (STEHELIN *et al.*, 1976). The *src* sequence in Rous sarcoma virus is related but not exactly identical to its progenitor, *protosrc*. There may be several unrelated *protosrc* sequences in normal chicken DNA. It is assumed that *protosrc* existed first in chicken DNA then was acquired rather recently by RNA-containing viruses. It is theoretically possible that *src* existed first in the virus and was introduced into the germline of chickens in the evolutionary past, only more recently becoming *protosrc*. Usually, however, *protosrc* is considered to be the original sequence, as its name implies.

Avian myeloblastosis virus, a leukosis virus, is also a descendant of RAV-0. The dashed line of Fig. 45 signifies that it is not known whether chickens carry a *"protoleuk"* sequence for forming leukosis viruses. It is also not known whether sarcoma and leukosis viruses originate independently or whether sarcoma viruses arise from leukemia viruses, as is suspected to be the case in laboratory-derived sarcoma viruses of mice (see later).

The second major group of RNA tumor viruses come from mice and, to a certain extent, from rats. The mouse viruses are an inordinately complex group. A large fraction of the mouse genome codes for virogenes (GILLESPIE, SAXINGER and GALLO, 1975; BENVENISTE *et al.*, 1977). Mice possess several different classes of virogenes, unrelated to each other (Fig. 46). Those retroviruses other than the type- C class will be discussed only briefly.

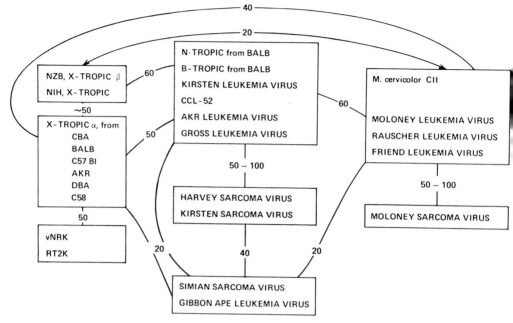

Fig. 46. Diagram of relatedness among mouse viruses

Type-A particles are intracellular and noninfectious, but they contain RNA physically similar to the RNA of the infectious RNA tumor viruses (YANG and WIVEL, 1973). They appear to be class 1 viruses coded by multiple genes (LUEDERS and KUFF, 1977). They are probably genetically distinct from type-B and type-C mouse viruses (WONG-STAAL et al., 1975). Mouse mammary tumor virus is a type-B RNA virus capable of causing breast tumors in mice. Different strains of mouse mammary tumor virus have different oncogenic potential, but all the isolates are highly related to one another. Mouse mammary tumor viruses are not related to other mouse viruses.

The remaining viruses we will consider are all type-C viruses. They are all genetically interrelated and are composed of leukemogenic viruses, sarcomagenic viruses, or their progenitors. These viruses are often categorized according to their host range. The "xenotropic" mouse viruses fail to grow on mouse cells but replicate in cells of other species (LEVY, 1973). The "N-tropic" and "B-tropic" mouse viruses grow on some mouse cells but not others (HARTLEY, ROWE and HUEBNER, 1970; PINCUS, HARTLEY and ROWE, 1971). Specifically, N-tropic viruses grow only in mouse cells with an Fv-1^n allele (e.g. NIH Swiss mouse cells) and B-tropic viruses grow only in mouse cells with an Fv-1^b allele (e.g. BALBc mouse cells). Other mouse viruses grow well in both types of mouse cells; those are called N, B-tropic. Finally, a newly-discovered group of viruses from wild mice, Mus musculus, (OFFICER et al., 1973) grow both in mouse cells and in cells from other animals. They have been called "amphotropic" (RASHEED, GARDNER and CHAN, 1976; HARTLEY and ROWE, 1976). The N, B-tropic and the amphotropic viruses are of special interest because they are tumorigenic; the N, B-tropic viruses provide laboratory models for the formation of oncogenic viruses while the amphotropic viruses are the natural examples for testing the models.

It is exceedingly difficult to deduce progenitor-descendant relationships among the mouse viruses because their hosts are highly inbred. Thus, endogenous mouse viruses carry genomes that bear more homology to DNA from the strain of laboratory mice that yielded them than to DNA from other strains of mice (CALLAHAN et al., 1974). This is in spite of the virtual identity in the bulk of the unique sequence DNA among mouse strains and subspecies (RICE and STRAUSS, 1973). Thus, in inbred mice the virogene sequences have evolved at an unusually rapid rate and their transcripts can become viruses that are *indigenous* to mouse strains, but not *endogenous* to mice. CALLAHAN et al., (1974) ascribe this unusually rapid rate of virogene evolution in laboratory mice to virus activation and reintegration, according to the protovirus concept (TEMIN, 1971). This notion is supported by the finding that an endogenous virus from *Mus cervicolor*, a wild asian mouse, carries a genome that shows essentially the same homology to genes in *Mus caroli* and *Mus musculus* as is exhibited by the unique sequence DNA of *Mus cervicolor*.

Nevertheless, it is valuable to compare the relatedness among the indigenous mouse viruses (Fig. 46). There are two classes of *xenotropic* viruses, termed alpha and beta. They are considered separate classes because viruses of the alpha group exhibit more internal homology than homology to viruses of the beta group and *vice versa* (CALLAHAN et al., 1974). Nonetheless, they are considered to have arisen from divergent copies of a single virogene that was amplified in the past. The xenotropic viruses are related to the N-tropic, B-tropic and N, B-tropic viruses, collectively called the *ecotropic* mouse RNA viruses.

The xenotropic and most of the N-tropic and B-tropic viruses are not readily produced by mouse cells. They have been induced from cultured cells with chemicals (LOWY et al., 1971; AARONSON, TODARO and SCOLNICK, 1971) or transmitted from leukemic mice (GROSS, 1951; ROWE, 1972). By molecular hybridization these viruses are indigenous to the strain of mice from which they were obtained.

N, B-tropic viruses grow readily on all mouse cells where the state of the allele at the Fv-1 locus is Fv-1^n or Fv-1^b. Mouse cell hybrids of the constitution Fv-1^b/Fv-1^n can be produced and they are refractory to infection by N, B-tropic viruses as well as to infection by N-tropic and B-tropic viruses. Otherwise, N, B-tropic viruses grow ubiquitously on mouse cells. Ordinarily, they grow poorly on cells from other species. The N, B-tropic viruses can be generated in either of two ways, operationally; they can be generated by forced passage of B-tropic viruses through restrictive cells carrying the Fv-1^n allele and they can be isolated as leukemia viruses from mouse tumors (FRIEND, 1957; MOLONEY, 1960; RAUSCHER, 1962). Since all the oncogenic N,B-tropic leukemia viruses were isolated after repeated passages through animals, they, too, could represent forced passage through restrictive cells.

The mechanism of action of the product of the Fv-1 allele is still unknown. The genomes of N-tropic and B-tropic viruses differ (FALLER and HOPKINS, 1977) at many places (FALLER and HOPKINS, 1978). N,B-tropic viruses formed by forced passage contain RNA genomes that differ from the B-tropic parent, at least near the 5' end and possibly elsewhere (FALLER and HOPKINS, 1978). Thus, the Fv-1 allele seems to respond to alterations in the coding properties of the virus genome. As described earlier, the product of the Fv-1 locus stops infection prior to provirus integration and its action can be abrogated by prior infection with one virus particle. A possible explanation to account for the properties of this dominant gene is a protein produced in limited quantity (under 50 copies per cell) which binds tightly to proviral DNA, preventing its interaction with the host chromosome. It is difficult to imagine a normal role for such a protein, however.

The viruses isolated from spontaneous tumors comprise some of the most widely-used leukemia viruses of mice, the Gross-type leukemia viruses and the Friend-Moloney-Rauscher leukemia viruses. Each virus group is represented by several members that are more related to one another than they are to viruses of the other group.

The viruses of the Friend-Moloney-Rauscher group of viruses are characterized by complicated natural histories. FRIEND isolated the first virus of the group in 1957. While passing an Ehrlich ascites tumor in NIH Swiss mice she noticed a case of leukemia and was able to isolate a cell-free agent capable of causing leukemia in recipient mice. The participation of the ascites cells in the isolation of the virus is seldom mentioned, but to our knowledge it is not clear whether the virus came from the Ehrlich tumor or the carrier mouse or both. Until recently the question was academic, but since recent evidence suggests that tumorigenic potential arises from recombinations between nontumorigenic viruses and cell genes, it may be prudent to reemphasize the complication in the natural history of the virus.

In 1960, MOLONEY isolated a virus from a sarcoma called sarcoma 37 of BALB/c mice. Initially, the virus had very little tumorigenic activity but after repeated passages in BALB/c mice the leukemogenic titer increased. The repeated passaging to enhance the leukemogenic phenotype is reminiscent of the history of the Friend leukemia virus.

Rauscher mouse leukemia virus has an equally complicated natural history. Following the report of GROSS (1951) that AK mice, a strain of mice with a high incidence of "spontaneous" leukemia, contain an agent that causes leukemia in recipients, SCHOOLMAN and his colleagues (SCHOOLMAN et al., 1957) caged an AKR mouse with an NIH Swiss mouse and obtained a "spontaneous" lymphoblastoma in the NIH Swiss mouse. These tumors could be transmitted to recipient NIH Swiss or DBA mice with cell-free extracts.

RAUSCHER obtained a sample of a cell-free extract from one of the mice with lymphoblastomas, but the potency of the extract had decayed and was only 20% of that originally reported by SCHOOLMAN et al. (1957). Consequently, RAUSCHER (1962) used the forced passaging procedure that had been successful in increasing the potency of other RNA tumor viruses. Finally, after passages in BALB/c mice a mouse developed a tumor at the site of inoculation. The virus isolated from this tumor caused leukemia in recipient animals.

Both RAUSCHER and MOLONEY were experimenting on the Friend virus when they obtained their isolates. Moreover, the Rauscher virus was clearly isolated from mice that had received the AKR Gross-type virus. This led GROSS (1966) to speculate that the Rauscher virus and possibly the Moloney virus were in actuality mixtures of the Friend virus and the true leukemia virus, the Gross virus.

If xenotropic viruses are progenitors of leukemia viruses (see Section II on Origin of Retroviruses), it appears that leukemia viruses are progenitors of sarcoma viruses. We have already seen in the avian virus system that the nondefective avian sarcoma virus is a modified cell gene added onto a viral genome. The recipient virus is unidentified in this case; it could either be a leukemia or an endogenous virus. In the case of the mouse viruses SCOLNICK and his associates have accumulated evidence that the recipient virus is leukemogenic and that the cell gene has special properties. Before examining this evidence let us first review the natural histories of a few of the mouse sarcoma viruses.

In the early 1960's, HARVEY was working with the Moloney strain of mouse leukemia virus, passaging it in rats. She reported in 1964, that one of the rats developed a tumor which yielded a virus capable of causing tumors in rats and mice. Two years later, in 1966, MOLONEY observed sarcomas in mice after injecting them with massive doses of Moloney leukemia virus. The sarcomas yielded a virus capable of causing sarcomas when administered in low doses. HARVEY obtained a sarcoma virus by passing a leukemia virus in a low dose in a new animal species while MOLONEY achieved the same result by finding a rare sarcoma-causing virus in the leukemia virus stock. Probably, recombinational events were responsible in both instances for creating the sarcomagenic phenotype.

In 1967, KIRSTEN and his colleagues isolated a murine erythroblastosis virus from a strain of mice that had previously yielded Gross-type viruses (KIRSTEN et al., 1967). That same year KIRSTEN and MAYER reported the isolation of a sarcoma virus following the repeated inoculation of the erythroblastosis virus into rats. With each passage the virus became more virulent, finally becoming a potent sarcomagenic virus.

We are going to pay some attention to the facts that are known about the mechanism of formation of mouse sarcoma viruses because the topic is leading in the direction of a general understanding of how retroviruses cause cancer. We focus on the mouse viruses because more details have been obtained from them than from the avian viruses, but for perspective we shall review salient facts from the avian virus system.

In 1971, in his protovirus model TEMIN proposed that an endogenous, non-tumorigenic type-C RNA virus becomes a tumorigenic RNA tumor virus by genetic change caused in part by repeated recombination with cell genomes. It has become fashionable to equate the active portion of the cell genome with virogenes.

HAYWARD and HANAFUSA (1975) studied RAV-60, a virus formed by passing the exogenous virus, RAV-2 through uninfected chicken cells. They found that RAV-60 was related both to RAV-2 and RAV-0 but identical to neither. The genome of RAV-60 was more highly related to the RNA of RAV-0, the true endogenous chicken virus than it was to the RNA of RAV-2. HAYWARD and HANAFUSA concluded that RAV-60 was a recombinant virus, recombinant between the incoming RAV-2 and the RAV-0 virogene. The possibility that RAV-60 was a mixture of complete RAV-2 and RAV-0 viruses was discounted because the RAV-60 stock was purified by endpoint dilution on quail cells, a cell line that does not support the growth of RAV-2.

The following year SHOYAB, DASTOOR and BALUDA (1976) obtained evidence by molecular hybridization that the proviruses of avian myeloblastosis virus integrated next to the endogenous RAV-0 gene. The difficulty with the experiment is that it rests on quantitative differences from competitive molecular hybridization to distinguish among six related models for integration and consequently is not generally accepted as conclusive.

In 1977, KESHET and TEMIN reported that RAV-61 was a recombinant between the infecting high titer Bryan strain of Rous sarcoma virus and genes in normal pheasants (see HANAFUSA and HANAFUSA, 1973, for origin of RAV-61). The pheasant sequences acquired by RAV-61 were not related to the genomes of the reticuloendotheliosis virus, pheasant viruses, or other avian viruses (KESHET and TEMIN, 1977).

These results suggest that infecting RNA tumor viruses can recombine with host virogenes, giving rise to new, recombinant forms. The new forms can acquire new biological properties. Of special interest would be the properties of leukemogenesis and sarcomagenesis. With this in mind we turn to a discussion of the mouse sarcoma viruses.

The known sarcoma viruses of mice are all replication-defective; they are produced by cells only when the cells are coinfected by a helper virus capable of supplying the missing growth functions. In the cases cited above—Harvey, Moloney, and Kirsten sarcoma viruses—the helper virus was a leukemia virus and was present in the final virus stocks possessing sarcomagenic potential.

What is the nature of the sarcoma virus in mice? Is it a helper virus with a *src* gene added, as in the replication-competent avian sarcoma viruses? Apparently not, for the RNA genome of the sarcoma virus is actually smaller than the genome of the original leukemia virus (MAISEL et al., 1973; 1978). This difference in size not only prompted the idea that mouse sarcoma viruses *lacked* sequences present in the leukemia virus genome, it also provided a means for studying the sarcoma component in the absence of the RNA of the helper virus.

Since 1973, SCOLNICK and his colleagues have been accumulating evidence that the mouse sarcoma viruses are recombinants between the leukemia virus and genes in the DNA of the host cell. During the recombination event(s) cell sequences are added to the virus and at the same time viral sequences are deleted. The resultant virus acquires sarcomagenic potential and loses replication capacity.

SCOLNICK's is a particularly attractive model, but it is important to review the nature of the experiments on which the conclusions rest, for the early work had serious interpretive limitations.

In 1973, SCOLNICK et al. showed that Kirsten virus carried rat sequences in addition to sequences corresponding to the mouse leukemia virus genome. They obtained plasma from one of the original virus-shedding rats. The virus had never been grown in tissue culture. With this stock of Kirsten sarcoma virus they produced transformed nonproducer NIH Swiss mouse cells. These cells, but not the original uninfected NIH Swiss mouse cells, contained RNA related to an endogenous rat virus. SCOLNICK et al. (1973) stressed the interpretation that the rat virus-related RNA was the expression of the Kirsten sarcoma virus RNA in the nonproducer cells rather than the alternative that both the uninfected mouse cells and the infected transformed mouse cells carried the capacity to code for the rat virus RNA but only the transformed cells expressed the appropriate sequences.

SCOLNICK *et al.* (1973) also showed that DNA copies of the RNAs of the sarcoma plus leukemia virus, but not the leukemia virus alone, hybridized to RNA from cells producing the rat endogenous virus. They concluded again that Kirsten sarcoma virus carried RNA related to the genome of the rat virus. Of course, since the stock containing the sarcoma virus came through rats it could have consisted of *three* viruses, the Kirsten sarcoma virus, the helper erythroblastosis virus, and the rat endogenous virus.

Nevertheless, SCOLNICK *et al.* (1973) took the following tack, and subsequent work has shown that they are probably correct:

"During passage of Ki-MuLV in rats, a recombinational event occurred with information contained in rat cells. The process resulted in the formation of a recombinant between Ki-MuLV and sequences in the rat cells. Thus Ki-SV now contains some MuLV information plus additional information. Part of this additional information clearly has homology with information contained in rat type-C viruses. This rat information can be transduced into mouse cells during the process of transmission of Ki-MuSV from rat cells to mouse cells or mouse cells to mouse cells. Presumably, Mo-SV (Moloney sarcoma virus) resulted from similar events in mouse cells; different strains of Mo-SV may then simply reflect varying amounts or kinds of MuLV sequences associated with the transforming information."

It obviously followed that Harvey sarcoma virus should also possess rat virus-related sequences, since it was obtained by repeated passage through rats (HARVEY, 1964). SCOLNICK and PARKS (1974) showed this to be the case. The ratlike sequences were absent from the helper Moloney leukemia virus and appeared to be the same or similar to the ratlike sequences in Kirsten sarcoma virus. The types of experiments characterizing Harvey sarcoma virus were the same as those used to characterize Kirsten sarcoma virus, subject to the same alternate explanations.

Moloney sarcoma virus lacks these ratlike sequences (SCOLNICK and PARKS, 1974; SCOLNICK *et al.*, 1975) as would be expected since the virus was contained within mice throughout its natural history (MOLONEY, 1960). Apparently, the *src* sequences of Moloney sarcoma virus originated in mice (SCOLNICK *et al.*, 1975; FRANKEL and FISCHINGER, 1977) and were introduced into the sarcoma virus by recombination.

More recently, the data supporting a recombinational origin for mouse sarcoma viruses has been strengthened. ROY-BURMAN and KLEMENT (1975) showed that the RNA of Kirsten virus hybridized to DNA from rats, eliminating the problem of coding capacity *vs.* expression. Unfortunately, ROY-BURMAN and KLEMENT did not purify the sarcoma-specific RNA but relied on assays indicating a high biological titer of sarcoma virus to attest to the relative absence of leukemia virus and rat endogenous virus.

MAISEL, SCOLNICK and DUESBERG (1975) purified the small sarcoma virus subunit of Harvey sarcoma virus RNA, freeing it from the larger helper virus RNA, and showed that the small subunit carried the ratlike sequences. SHIH *et al.* (1978) purified Kirsten sarcoma virus RNA and mapped T1 RNase-resistant oligonucleotides specific for either the leukemia virus or the rat endogenous virus, combining the poly (A) selection technique of WANG *et al.* (1975) with the hybridization of oligonucleotides developed by COFFIN and BILLETER (1976). Within the

6000 nucleotide RNA molecule corresponding to the sarcoma virus, reading from the poly (A)-containing 3′ end, nucleotides 1—1000 come from the leukemia virus, nucleotides 1000—5900 (approx.) come from the rat, and nucleotides 5900—6000 come from the leukemia virus.

By the RNA fingerprinting method the ratlike sequences of Kirsten sarcoma virus are not distinguishable from sequences in rat endogenous virus, hence not distinguishable from a gene(s) in rats, but they are distinguishable from the ratlike sequences in Harvey sarcoma virus (SHIH et al., 1978). Oddly, the hybrid formed between the RNA of Kirsten sarcoma virus and rat DNA has a low thermal stability (our unpublished data).

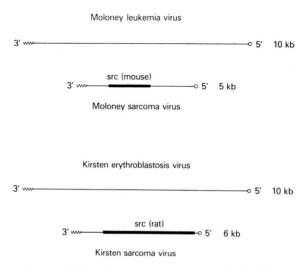

Fig. 47. Map of two mouse sarcoma virus RNAs.
Kirsten sarcoma virus and Moloney sarcoma virus are apparently formed by recombination between leukemia virus genomes (thin lines) and a particular cell gene or genes (thick lines). In both cases the cell genes replace the center of the leukemia virus genome. Each leukemia virus genome is 10 kilobases long while the sarcoma virus RNAs are only 5—6 kilobases in length

A simple view concerning the sequence arrangement in RNA of sarcoma viruses and the mechanism of their origin is diagrammed in Figs. 47 and 48. RNA from Moloney sarcoma virus is a 1500 nucleotide *src* sequence derived from mouse DNA flanked by 1000 nucleotides of Moloney leukemia virus sequence toward the 3′ end and 2500 nucleotides of Moloney leukemia virus sequence toward the 5′ end. The RNA of Kirsten sarcoma virus is arranged similarly with a 5000 nucleotide *src* sequence derived from rat DNA flanked toward the 3′ end by 1000 nucleotides of Kirsten erythroblastosis virus sequence and at the 5′ terminus by a very short piece of the erythroblastosis RNA. The positions of the leukemia virus sequences in the sarcoma virus RNA are the same as their position in the leukemia virus RNA, as far as can be determined; the 3′ end of the leukemia virus RNA ends up as the 3′ end of the sarcoma virus genome while the 5′ end of the helper is recovered at the 5′ end of the sarcoma virus RNA.

SHIH *et al.* (1978) proposed that the recombination event creating these hybrid RNAs is an extrachromasomal, copy-choice mechanism involving reverse transcription, as deduced earlier by WEISS, MASON and VOGT (1973) in the avian virus system. Early in the infection of a cell by a leukemia virus a DNA provirus is made from the invading viral RNA, using reverse transcriptase. If the viral RNA is aligned at the time with a molecule of endogenous virus RNA, then the leukemia virus reverse transcriptase can begin by copying the leukemia virus RNA, switch to the endogenous virus RNA at some point, then switch back to the leukemia virus RNA to finish up (Fig. 48). This model leads to an insertion of a cell sequence

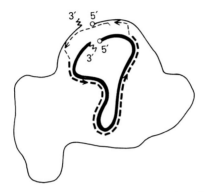

Fig. 48. Copy-Choice model for sarcoma virus formation.
The leukemia virus RNA (thin line) "aligns" with an endogenous "type-C RNA" (thick line). During copying of the leukemia virus RNA into DNA (see Fig. 5) the DNA polymerase switches, templates and copies the type C RNA. Before finishing, the polymerase switches back and finishes the leukemia virus RNA. The position of the second switch seems to be more random than the position of the first switch. After SHIH *et al.* (1978)

within the viral genome, deleting the viral genome in the process. It does not necessarily alter the base sequence of the RNA regions that become incorporated into the sarcoma virus genome.

   The provirus formed by copy-choice can then integrate into the host cell as the leukemia virus would, but the integration would not lead to virus production except in cases where a leukemia virus integrated into the same cell. The sarcoma virus should have two functions derived from the leukemia virus parent. It should carry signals requisite for initiating reverse transcription. Presumably the termini perform this function and will be conserved in replication-defective sarcoma viruses, as TRONICK *et al.* (1978) have found. TRONICK *et al.* also suggest that the termini would be required for forming the dimer structure of the viral RNA or for providing leader sequences. The other function that must be transmitted to the sarcoma provirus is the ability to integrate into the cell genome. It is unlikely that cell DNA integrates with the efficiency of proviruses; some special sequence is probably required. Again, the ends of the RNA have been implicated in the process of integration as well as in the other functions described above. However, in all sarcoma viruses known to us about 1000 nucleotides of helper virus enters the sarcoma virus genome. In the avian viruses the host range marker maps in this region. Possibly a substantial portion of the 3′ end of the RNA codes for integration functions.

What is the nature of the new sequences acquired from the rat, in the cases of Kirsten and Harvey sarcoma viruses, or from the mouse, in the case of Moloney sarcoma virus? SCOLNICK, GOLDBERG and WILLIAMS (1976) distinguished between "type-C rat genes" and "rat cellular genes" a nomenclature equivalent to the "virogenes" vs. "other cell genes" we use in this monograph. SCOLNICK, GOLDBERG and WILLIAMS concluded that the new sequences in Kirsten and Harvey sarcoma viruses were rat virogene sequences because their expression can be regulated by bromodeoxyuridine induction, because they can be copied by reverse transcriptase and because they are present as multiple-copy elements in the rat genome.

SCOLNICK, MARYAK and PARKS (1974) pointed out, however, that there is more than one rat virogene. Two different cell lines derived from Osborne-Mendel rats produced different viruses. The rat thymus cell line, RT 21c, releases a leukemogenic virus, RT 21c virus (CREMER et al., 1970). The rat kidney cell line, NRK, releases an RNA virus, vNRK (DUC-NGUYEN et al., 1966). Both viruses are ecotropic; they grow only in rat cells (SCOLNICK, MARYAK and PARKS, 1974). The ratlike sequences in Kirsten and Harvey sarcoma viruses are related to vNRK, not to RT 21c. The vNRK and RT21c virogenes are independently regulated. The product of the RT 21c virogene is a 34—35 S RNA; that of the vNRK virogene is somewhat smaller (SCOLNICK et al., 1976).

ANDERSON and ROBBINS (1976) also found homology between the rat-originated components of Kirsten sarcoma virus and the virus from NRK cells of Osborne-Mendel rats. They saw no or little homology with the leukemia virus of Wistar-Furth rats, Jones chloroma rat virus or Fisher rat leukemia virus.

Amphotropic mouse viruses, mouse leukemia viruses and endogenous mouse viruses can package the vNRK virogene transcripts as they, themselves, replicate in rat cells (SCOLNICK et al., 1973, 1976; ANDERSON and ROBBINS, 1976). During this "rescue" of the virogene transcripts by the infecting viruses as much as half of the RNA recovered from extracellular virus particles is vNRK RNA (SCOLNICK et al., 1976), though it is not certain whether the vNRK RNA is packaged by the infecting virus or by rat "particles".

Apparently, infecting type-C viruses *specifically* package transcripts of those virogenes potentially coding for type-C viruses (GOLDBERG et al., 1976). Hemoglobin message RNA is packaged only rarely, if at all. The RNA virogenes that code for a type B virus, mouse mammary tumor virus, and the RNA of a provirus coding for a type-D virus, the Mason-Pfizer monkey virus, are not rescued by infecting type-C viruses.

Rats are not the only animals harboring RNA that can be rescued by RNA tumor viruses. HOWK et al., (1978), identified a 30S RNA in mouse cells that could be rescued by superinfection. The 30S RNA rescued by Moloney leukemia virus and separated from the 38S RNA of the leukemia virus was unrelated to the Moloney leukemia virus genome and was unrelated to the RNA of mouse xenotropic and amphotropic viruses. The 30S mouse RNA is coded by multiple genes and is inducible with bromodeoxyuridine (HOWK et al., 1978). No mention was made concerning whether the 30S RNA was homologous to the *src* sequences of Moloney sarcoma virus.

SHERWIN *et al.* (1978) reported on a similar RNA species in mouse cells. They found that baboon endogenous virus could rescue a 30S RNA when replicating in mouse cells. They also learned that the RNA was coded by multiple genes in mouse DNA and that it was not related to any known mouse virus. It was related, however, to abundant RNA molecules produced by several mouse strains and species.

Other workers reported on RNA species in mouse cells having some properties of the 30S RNA, but where identity is less certain. TSUCHIDA and GREEN (1974) discovered a 26—28S RNA in normal mouse cells with homology to Moloney sarcoma virus. MUKHERJEE and MOBRY (1975) found an RNA in proliferating mouse cells with homology to the genome of an endogenous mouse virus, the S_2Cl_3 virus. This RNA was more abundant in embryos than in newborns and more in reproductive and proliferating tissues of adults than in other adult tissues. Finally, GETZ *et al.* (1977) reported higher levels of RNA homologous to the AKR virus in chemically-transformed cells than in normal control cells.

Other cells produce "rescuable" RNA. Mink cells produce an RNA that can be rescued by Kirsten sarcoma virus (SHERR, BENVENISTE and TODARO, 1978) and probably by mouse amphotropic viruses (CHATTOPADHYAY, 1978). It is not known whether the RNA is related to that of an endogenous mink virus described by BARBACID, TRONICK and AARONSON (1978). The RNA of the endogenous cat virus, RD114, is found at an elevated level in many types of tissues, including some tumors (NIMAN *et al.*, 1977).

We do not know whether the endogenous type-C RNAs of mice, mink, or cats participate in the origin of sarcomagenic (or leukemogenic) viruses in their respective species. We do not know whether these RNA molecules have a positive role in the development or maintainance of the animal. Nonetheless, it is probably time to suspect that SCOLNICK's model for the formation of mouse sarcoma viruses can be generalized to other mammalian species and it is probably time to be concerned about its participation in natural forms of cancer.

SCOLNICK's model can be extended to other viruses of mice, for there are now several examples of tumorigenic recombinant retroviruses. The spleen focus-forming virus component of the Friend leukemia virus is a laboratory recombinant between an ecotropic and a xenotropic virus (TROXLER *et al.*, 1977), one crossover being near the *env* gene, some 1000—1500 nucleotides from the 3' end of the RNA genome of the virus (TROXLER *et al.*, 1977). The amphotropic MCF viruses are also laboratory recombinants between ecotropic and xenotropic mouse viruses at the envelope gene (ELDER *et al.*, 1977). The MCF virus appears late in the pre-leukemic period in AKR mice (KAWASHIMA *et al.*, 1976) and may play a role in the etiology of spontaneous lymphomas of certain laboratory mice, according to HARTLEY *et al.* (1977). Certainly, the avian sarcoma viruses are recombinants between a cell gene and an avian virus (STEHELIN *et al.*, 1976).

The fact that tumorigenic viruses are recombinants may not be restricted to laboratory viruses. The lymphoma viruses recovered from wild mice have properties of both xenotropic and ecotropic viruses, being amphotropic in host range (RASHEED, GARDNER and CHAN, 1976; HARTLEY and ROWE, 1976). Unlike the MCF virus, which is neutralized by antisera directed against either the xenotropic or the ecotropic mouse viruses, the wild amphotropes are neutralized by neither serum. Nevertheless, it may be that the natural viruses of mice, too, are recombinants.

The feline leukemia virus may also be a recombinant virus. The simplest explanation of the data pertaining to its origin is that it was a rat virus that infected cats and recombined repeatedly with cat DNA, becoming tumorigenic in the process (see pp. 64—68). This virus is the etiologic agent for leukemia and lymphomas in cats (see pp. 15, 85). Cats possess sequences in their DNA that are related to the genome of feline leukemia virus (BENVENISTE, SHERR and TODARO, 1975). Tumors of some leukemic cats, especially virus-shedding tumors, possess new DNA sequences related or identical to feline leukemia virus genes, but other tumors, especially in virus-negative cats, apparently lack new viral sequences at a detectable level (KOSHY et al., 1979). Stipulating that tumors in virus-negative cats are nonetheless caused by feline leukemia virus, it follows that little, if any, of the viral genome need be integrated into DNA of the animal to produce a tumor.

RNA tumor viruses may act chiefly by activating and modifying cell genes that are used for normal growth, development and differentiation. They may act as "site-specific" mutagens, having some homology with the genes they affect. This feature would make them more potent than chemical mutagens, which must act randomly. Nevertheless, there is probably a randomness to natural mechanisms for viral-induced cancer, concerning which part of the virus genome integrates, where in the chromosome it integrates, and the physiological state of the infected cell. We are especially attracted to a model where a retrovirus productively infects a cell and acquires a host repeated DNA sequence. This sequence might permit the virus to move around the host chromosome, integrating at many alternate sites. Depending on the virus titer in the animal, there may be a significant probability of one of those secondary infections arresting the differentiation of an embryonic, fetal, stem, or incompletely-differentiated cell, causing cancer.

This scheme ignores the physiology and phenotype of cancer cells. Probably, the physiologies of different tumors differ fundamentally as their phenotypes are known to differ. To treat cancer as a single disease at the phenotypic level seems hopeless. However, there is still the possibility that the varied phenotypes of the cancer cell are manifestations of a genetic change whose nature is common to all tumors. It is in the direction of elucidating this change, whether it is a change in DNA as TEMIN (1971) proposed or an alteration in its expression as RNA as HUEBNER and TODARO (1969) favored or some alternate abnormality in genetic control, that fundamental inroads toward understanding the causes of cancer will be made. In the process, it is likely that we shall come to understand aspects of development and evolution that are now obscure.

VII. Human Retroviruses

1. The Logical Problem

Several lines of evidence suggest an association between RNA tumor viruses and cancer in man (GILLESPIE, SAXINGER and GALLO, 1975; VIOLA et al., 1976; GALLO and GILLESPIE, 1977). If the link is a causal one, the availability of the causative agent could be useful both for research and for therapy. The rare isolation of infectious type-C RNA viruses from cultured human tissues has been reported by several groups (see later). These reports have generated criticisms of two major forms: (1) If type-C RNA viruses cause human cancer, they should be regularly isolated from the neoplastic tissue and (2) the viruses could be contaminants which were inadvertently introduced into the tissue sample during the course of laboratory manipulation.

As our knowledge increases concerning the relationship between RNA tumor viruses and cancer in animals, it is becoming clear that the first criticism is unwarranted. In cats, for example, feline leukemia virus (FeLV) is one causative agent for leukemias and lymphosarcomas (BRODY et al, 1969; JARRET et al., 1973). FeLV consists of viruses that are freely transmissible among cats and many cats excrete appreciable levels of infectious virus. However, in some leukemic cats infectious, replicating virus cannot be detected. In addition, FeLV viral proteins, proviral DNA and antibodies against viral proteins are missing in many of these cats. These virus-negative cats do carry a tumor-specific antigen (FOCMA) whose production is in some way associated with virus infection. Therefore, in cats, one would not expect to isolate replicating virus in every case. Isolation of infectious virus from animal neoplasms varies widely from animal to animal. The isolation of infectious virus from leukemic gibbons is relatively easy. The isolation of virus from leukemic cows is rather difficult. Infectious virus has never been isolated from leukemic dogs. Infectious virus is rarely isolated from mice or chickens with spontaneous leukemia.

The ease of virus isolation will probably vary with different animals and even with different neoplasms in the same species. A priori one could not have predicted where to place humans in such a matrix. With available data and technology, it is now apparent that the isolation of infectious virus from humans will occur extremely rarely.

The second criticism, that of laboratory contamination, is a more difficult problem. The criticism is especially apropos in the cases of existing candidate human RNA tumor viruses for all reported isolates are very closely related, though not identical to laboratory isolates from nonhumans.

If the transmission of type-C RNA viruses to humans has occurred, one might expect the transmitted viruses to be related to, but easily distinguishable from other viruses obtained from the progenitor animal. In fact, the viruses obtained from cultured human cells are not easily distinguished from laboratory viruses, prompting the conclusion by some that they are laboratory contaminants. However, if the argument is taken to its conclusion, it states that viruses recently transmitted from one species to another will very closely resemble other viruses from the progenitor animal and in this context the close relationship of the candidate human viruses to viruses obtained from nonhumans might have been predicted.

There are several approaches that can be used to dispel the criticism of laboratory contamination. One is to obtain evidence for proteins or nucleic acids of the particular virus type in fresh, uncultured human tissue. A second is to repeatedly isolate virus from tissues of the same patient, several times from the same tissue or, preferably, from different, independently-drawn tissue samples. A third is to repeatedly isolate the same virus type from several patients under different experimental conditions. A fourth approach would be, having a candidate human virus, to isolate DNA from fresh, uncultured tissues of the same patient and to use it for transfection. If transfection were accomplished, and if the transfected virus were indistinguishable from the candidate virus obtained by conventional cocultivation, there should no longer be a question of laboratory contamination. This experiment has not yet been done, but perhaps it would be a useful one in the future.

In the absence of a conclusive experiment it is necessary to weigh existing circumstantial evidence. The discussions that follow have that as their objective.

2. The Natural History of Primate, Type-C, RNA Viruses

This section will be a historical account of the isolation of viruses relevant to the topic of candidate human viruses. The nonhuman primate, type-C RNA viruses consist of two major groups, one group is infectious for but not endogenous to primates, and the second group is endogenous to baboons. The first group consists of simian sarcoma and associated virus and gibbon ape leukemia viruses and is clearly tumorigenic in animals. The second group consists of viruses isolated from several species of baboons. The baboon viruses under experimental conditions so far utilized do not cause cancer. Some details of nonhuman, type-C RNA viruses are presented in Table 3.

Simian sarcoma virus and *simian sarcoma-associated virus (SSV, SSAV)* is a virus complex consisting of members that can be separated by biological properties into at least two members: a sarcoma-causing, nonreplicating agent (simian sarcoma virus or SSV) and a fully replication-competent virus not capable of causing tumors (simian sarcoma-associated virus or SSAV). Usually the ratio of SSAV : SSV in the complex exceeds 100 : 1 or 1000 : 1. It is not clear whether SSV and SSAV represent the only two components of the complex or reflect ends of a spectrum whose interior consists of viruses defective in particlar functions. Less than 1% of the virus particles in a typical preparation of SSV-SSAV are infectious. Both cDNA and RNA of SSV-SSAV are comprised 30% or more of sequences not found in SSAV. For simplicity and to be consistent with our previous publications concerning this virus, we will refer to the complex simply as SiSV.

SiSV was isolated from a fibrosarcoma of a woolly monkey (see p. 51 and Fig. 30). The monkey was trapped in South America as an infant and was housed in a pet store for several weeks but for most of its life was a household pet. The monkey was kept in association with other animals, one of these was a gibbon ape. At 3 years of age the monkey developed a fibrosarcoma. Type-C virus was observed in the tumor by THEILEN et al. (1971) and virus was isolated from extracts of the tumor tissue by WOLFE et al. (1971). Viruses were isolated from

Table 3. *Summary of primate and candidate human, type-C RNA viruses*

Virus	Origin	Common host cells used for propagation[a]
Nonhuman primate viruses		
SiSV	woolly monkey fibrosarcoma	71 AP 1, NRK, NC 37
GaLV-1	gibbon lymphosarcomatous tumor	UCD-144-MLA
GaLV-2	gibbon myeloblasts	NC 37
GaLV-3	gibbon lymphoblasts	cells of origin[b]
Gbr-1	normal gibbon brain (uninoculated)	A 204
Gbr-2	"normal" gibbon brain	FEC, Ib 1 Lu
Gbr-3	"normal" gibbon brain	DBS-FRhL-1, Tb 1 Lu
BaEV (M 28)	normal baboon testes	cells of origin
BaEV (M 7)	normal baboon placenta	A 7573[c], Tb 1 Lu, MV 1 Lu, A 204
BaEV (455 K)	normal baboon kidney	A 7573[c], Tb 1 Lu
BaEV (8 K)	normal baboon kidney	A 7573[c]
BaEV (BILN)	leukemic baboon inguinal lymph node	cells of origin[b]
BaEV *(P. papio)*		
BaEV *(T. gelada)*		
Candidate human viruses		
HL 23 V	human leukemic myeloblast culture (HL 23)	A 204, A 7573[c]
HL 23 V$_{bab}$	HL 23 V	HOS
HL 23 V$_{sisv}$	HL 23 V	Tb 1 Lu
HEL-12 V	normal human embryo lung (HEL-12)	cells of origin
HEL-1 V	normal human embryo lung (HEL-1)	cells of origin[b]
L 104 V	human lung tumor cells	XC
SAK 21 V	human leukemic marrow	NRK, R-970-5, SIRC

[a] See text for description of cell lines.

[b] Cells of origin = cells from which virus was obtained. Otherwise virus is produced following transmission to a suitable target cell.

[c] Also called FCF 2 Th.

several samples of the extract in different ways, giving rise to the several laboratory isolates now in use. SiSV causes fibromas or regressing, well-differentiated fibrosarcomas in primates (WOLFE et al., 1971).

In one case a cell-free extract of the fibrosarcoma was applied to cultured marmoset skin fibroblast cells (line 1283). The cells became infected 1283 (SiSV). The virus-producing cells were inoculated into a marmoset (marmoset number 71 AP1). The marmoset developed a tumor, a fibrosarcoma which was cultured. The cultured cells, 71 AP1 (SiSV) eventually produced SiSV (71 AP1). Early passage 71 AP1 (SiSV) cells do not produce virus. SiSV (71 AP1) has been used to infect many other cell lines. Two in common use are HF marmoset skin fibroblasts and NRK rat cells (not to be confused with KW23, see below).

In a second case, the extract was inoculated directly into a marmoset (marmoset number 71 Q1). The marmoset developed a fibrosarcoma. The tumor was put into culture and produced virus. This virus, SiSV (71 Q1), has been used to infect NC37 human lymphoid cells. The virus produced by the infected NC37 cells does not have transformation capabilities. Earlier, an NC37 (SiSV) line had been created using SiSV (71 AP1), but this line eventually stopped producing virus, and as far as we know, is no longer in use. This infected NC37 line also produced nontransforming virus.

In a third case the extract was placed upon NRK rat cells that had been previously transformed by Kirsten murine sarcoma virus. This line, KW23 or KNRK (SiSV), produced SiSV. From this virus population a nontransforming virus was isolated by endpoint dilution. This virus is called M55 or M55 (NRK).

The preparations of SiSV obtained in the different ways outlined above are indistinguishable using immunological techniques to compare some of the major structural proteins (gp70, p30 and p12). By the criterion of molecular hybridization, however, the genomes of SiSV (71 AP1), SiSV (NRK) or SiSV (KNRK), and SiSV (NC37) are all slightly different from one another (unpublished results with M. Reitz and R. G. Smith). Most notably, SiSV (NC37) lacks sequences found in other viruses. Whether this is related to its lack of transformation potential is not known.

SiSV produced by any given cell line may not consist of a collection of particles having RNA with identical nucleotide sequences (see Section IV on Organization of Infectious Retrovirus Genes). The proviral DNA sequences of the 71 AP1 (SiSV) cells consists of related, nonidentical sequences. Possibly, the RNA of the virus produced by these cells is heterogeneous also. It may be noteworthy that different cell lines producing SiSV were constructed without cloning the virus or the cells. The point is raised because if one attempts to compare a candidate human virus with known laboratory virus stocks, the experiments are more difficult to design if one or both of the viruses are heterogeneous. As far as we know, this is a possible complication with molecular hybridization analyses of virus genomes but not with comparative immunological studies of virus proteins.

The first isolate of gibbon ape leukemia virus (GaLV) came from a 3.5 year old female gibbon (Hylobates lar) having a spontaneous lymphosarcoma with tumors consisting of prolymphocytic cells in the lymph nodes, liver and bone marrow (KAWAKAMI et al., 1972). Tumor cells (UCD-144-MLA), primarily lymphoid, were established in culture, and they produced the virus. The gibbon was supplied by the San Francisco Medical Center and was a member of a group of six imported gibbons housed in a single cage. They had been exposed to radiation during the course of an aging study. A male cagemate of the virus-producing female gibbon also showed histological evidence of type-C virus particles. The virus from the female gibbon has been called GaLV-1, the San Francisco strain of GaLV or SLV. The UCD-144-MLA cells cause lymphocytic leukemia in gibbons. So far the purified virus does not cause neoplasms.

KAWAKAMI and BUCKLEY (1974) obtained the second isolate of GaLV from a gibbon *(Hylobates lar)* with a spontaneous granulocytic leukemia. The gibbon was a part of a large SEATO colony in Bangkok, Thailand, from which DE PAULI *et al.* (1973) had earlier reported 5 cases of "spontaneous" granulocytic leukemia. It may be noteworthy that the 5 leukemic gibbons were representatives of a group inoculated with blood of humans with malaria. The virus was transmitted from the gibbon tumor to human lymphoid NC37 cells, but the techniques used to accomplish this have not been reported. The second GALV isolates have been called GaLV-2, the SEATO or Thai strain of GaLV or SMV.

Three new isolates of GaLV were obtained from brains of "normal" gibbons (TODARO *et al.*, 1975). The animals were imported to Louisiana from Southeast Asia and could have been part of the Thailand SEATO colony. Two animals were inoculated intracerebrally with brain extracts from humans with kuru disease and the third, uninoculated was a cagemate. The inoculated animals died within four months of injection and tissues were frozen for 7 years, until 1975. The brain tissues were initially established in culture and later were cocultivated with uninfected cells from various animals. The virus produced by the target cells were termed GBr-1, GBr-2, and GBr-3 (see Table 3 for details).

The most recent GaLV isolate came from a seven year-old male gibbon *(Hylobates lar)* with spontaneous malignant lymphoma and lymphoblastic leukemia (GALLO *et al.*, 1978). The animal was housed on Hall's Island, Bermuda, between 1974 and 1976 as part of a free ranging colony. Previously, the gibbon was colonized in Madrid, Spain, but its earlier history is unknown. Shortly after diagnosis the animal was flown to Bethesda, Maryland, where it died a few days later. Examination of many tissues obtained at death revealed virus particles or biochemical markers in many tissues, both involved and uninvolved. Virus was produced by blood leukocytes established in cell culture and was transmitted to target cells from several other tissues by cocultivation.

All of the isolates of GaLV are related to one another and to SiSV. Serologically, their p30 proteins are indistinguishable and their gp70 and reverse transcriptases are highly related. They contain divergent p12 proteins. Their RNA genomes are related, but distinguishable. Because of their interrelatedness with one another and their wide divergence from most other type-C RNA viruses, excepting certain mouse viruses, we consider SiSV and GaLV to be a family of oncogenic, type-C RNA viruses infecting primates. The term "primate type-C RNA virus" is often used, but it should be noted that the use of the word "primate" in this terminology reflects the supporting host and not the animal that originated the virus. The progenitor animal for the SiSV-GaLV family was probably an ancestral mouse or rat (WONG-STAAL, GALLO and GILLESPIE, 1975; LIEBER *et al.*, 1975). This is in contrast to the virus isolated from baboons, also a primate type-C

RNA virus, wherein the baboon is probably the progenitor animal as well as the supporting host.

There are at least seven isolates of *baboon endogenous virus (BaEV)*, from seven different animals including four species of the genus, *Papio* and one member of the genus, *Theropithecus*. These viruses do not cause tumors in animals, and they do not transform fibroblasts in tissue culture.

Virus particles were first noticed in placentas of normal yellow baboons *(Papio cynocephalus)* (KALTER *et al.*, 1973; and LAPIN, 1973) and have since been seen in a variety of tissues from several baboon species. The first isolate of the baboon endogenous virus, BaEV, was called M28 and was obtained by infecting normal baboon testes cells (line 731) with feline sarcoma virus (MELNICK *et al.*, 1973 and BENVENISTE *et al.*, 1974). Because of the possible presence of the feline virus, M28 has not been routinely used.

The first isolate of BaEV free of other viruses was obtained by cocultivating placental cells of *Papio cynocephalus* with several uninfected target cells (BEN- VENISTE and TODARO, 1974). The first target cell to show evidence of viral production was a line of canine thymus cells (FCF2Th, also called A7573), and the virus produced was called M7. The virus was subsequently transmitted from the infected canine cell line to bat (line Tb1Lu) and mink (line MV1Lu) cells. All of these lines are in use. cDNA transcripts of M7 RNA hybridize fully to DNA from normal baboons and DNA from canine thymus cells producing M7 virus, but not to DNA from uninfected canine thymus cells (BENVENISTE and TODARO, 1974). This and many other similar experiments classify M7 as an endogenous virus of *P. cynocephalus*.

Four more isolates of BaEV were obtained from other baboons (TODARO *et al.*, 1974). Though not explicitly stated, it is our understanding that at least two of the animals (numbers 8 and 455) were dogface baboons, *Papio anubis;* baboons that are very closely related to *P. cynocephalus*. Bab-8K arose from kidney cells that had been cultured for 3 passages, treated with iododeoxyuridine, then cocultivated with canine thymus cells. Bab-455K arose from kidney cells that had been cultured for 3 passages, then cocultivated with canine thymus cells without prior treatment of the baboon cells with iododeoxyuridine. Another isolate of Bab-455K was obtained following iododeoxyuridine treatment of the baboon cells, but it is our understanding that this line is not in use.

One isolate of BaEV was obtained from the royal baboon *(Papio hamadryas)* (LAPIN, 1973). The mother of the animal involved was inoculated with the blood of a leukemic human. The inoculation program was followed by an "outbreak" of leukemia among the baboon colony. The inoculated mother was bred to an uninoculated male and the inguinal lymph node of the offspring was placed in tissue culture and produced virus. The virus is referred to as the BILN virus and is usually grown in its natural host in inguinal lymph node cells.

BaEV has also been isolated from the western baboon *(Papio papio)* and the gelada *(Theropithecus gelada)*.

The M7, 8K and 455K varieties of BaEV are virtually indistinguishable, measuring the antigenic relatedness of their major structural proteins using competitive radioimmune assays or probing the nucleotide sequence of their RNA genomes using molecular hybridization.

3. Candidate Human Retroviruses

Virus markers in human tumor cells, especially in blood leukocytes of leukemic patients have been regularly reported, starting in 1970 (GALLO, YANG and TING, 1970). Reverse transcriptase (BAXT, HEHLMAN and SPIEGELMAN, 1972; SARNGA-DHARAN et al., 1972) and viral-like RNA (KUFE, HEHLMAN and SPIEGELMAN, 1972; HEHLMAN, KUFE and SPIEGELMAN, 1972; GALLO et al., 1973; MILLER et al., 1974; LARSEN et al., 1975; TAVITIAN et al., 1976) in particles resembling RNA tumor viruses and capable of synthesizing cDNA with viral sequences (BAXT and SPIEGELMAN, 1972; BAXT, 1974; GALLO et al., 1973, 1974; MAK et al., 1974, 1975) have been described. Reports of viral proteins other than reverse transcriptase exist (STRAND and AUGUST, 1974; SHERR and TODARO, 1974, 1975; NOOTER et al., 1975, MELLORS and MELLORS, 1976; METZGAR et al., 1976), but remain unconfirm-ed. A recent literature describing the presence of antibodies in human sera directed against surface viral proteins is accumulating (AOKI et al., 1976; SNYDER et al., 1976; KURTH et al., 1977). However, the serological specificity of these antibodies is being investigated.

Antibodies which were thought to precipitate labeled virus apparently pre-cipitate labeled globulins from the calf serum used to grow the infected cells (AARONSON, personal communication). AARONSON suspects that humans raise the anti-globulin antibodies because of the milk they drink. Antibodies which were thought to be directed against purified viral gp70 are probably instead directed against a carbohydrate added posttranslationally by the host cell (BARBACID, BOLOGNESI, and AARONSON (1980). Antibodies in humans directed against viral reverse transcriptase (PROCHOWNIK and KIRSTEIN, 1977; JACQUEMIN, SAXINGER, and GALLO, 1978) remain uncontested.

The release of virus-like particles from blood leukocytes or marrow cells from leukemic patients has been described; (MAK et al., 1974; VOSIKA et al., 1975), but these particles are usually not infectious and may reflect release of intracyto-plasmic particles from degenerating cells.

Reports of the release of infectious viruses from human cells are rare. Below are described infectious viruses that have been released from human cells and are now or were at one time seriously considered to be candidate human RNA tumor viruses. Some details of the candidate human viruses are presented in Table 3.

(1) The ESP-1 virus. PRIORI et al. (1973) obtained a virus from pleural effusion cells from a 5-year old American child with Burkitt lymphoma (American type). The virus ESP-1 was a typical type-C particle, containing reverse transcriptase and viral interspecies determinants. Though the cultured Burkitt cells originally produced usable quantities of virus, virus production decreased with continued culturing and persistent mycoplasma infection developed. Transmission of ESP-1 to secondary uninfected cells failed. This, and the findings of murine-related antigens in the virus has discouraged further characterization of it. The history and current status of ESP-1 has been recently reviewed by DMOCHOWSKI and BOWEN (1978).

(2) The RD114 virus. MCCALLISTER et al. (1972) inoculated prenatal kittens with a cell line (RD) derived from a human with a rhabdomyosarcoma, producing tumors in the kittens. The RD cell line did not produce type-C virus particles, but

two of the cat tumors did, as did a cell line (RD114 cells) derived from one of the tumors. The cell line had a human karyotype. It was shown that the RD114 virus differed completely from known feline and mouse type-C virus, the basis for considering it as a candidate human virus. Later, it was shown that RD114 was a new feline virus, an endogenous type-C RNA virus of domestic cats (GILLESPIE et al., 1973; NEIMAN, 1973).

(3) *The HL23 viruses*. For several years no new candidate human type-C RNA viruses were reported, probably because of the unfortunate experiences with ESP-1 and RD114. GALLAGHER and GALLO (1975) reported the production of virus particles from blood leukocytes of a patient with acute myelogenous leukemia (AML), patient HL23, and they and their colleague (TEICH et al., 1975) transmitted this virus(es) (HL23V) to uninfected target cells wherein it was readily propagated. It soon developed that two viruses were present, one related or identical to SiSV and another related or identical to BaEV. Of several new viruses obtained from cancer patients, HL23V is the most exhaustively studied.

The HL23 patient was a 70-year old woman at the time she presented with symptoms of AML on October 16, 1973. On this date prior to the initiation of chemotherapy, a sample of blood leukocytes was initiated in tissue culture as part of a program involving the study of cells from many patients with different forms of leukemia (GALLAGHER et al., 1975). Of cells from 20 patients tested at the same time and 50 subsequently, only cells from patient HL23 has released virus. The first sample of cell released transmissible virus by passage 5 and by passage 10 virus was detectable in cultured cells themselves by electron microscopy and extracellular reverse transcriptase assays. Cells frozen at passage 2, before transmissible virus was detectable, were reinitiated in culture and by passage 10, they, too, produced virus detectable by electron microscopy.

A portion of the fresh, uncultured blood leukocytes received in October, 1973 was utilized for biochemistry. A reverse transcriptase activity was identified in these cells and shown to be antigenically related to the reverse transcriptase of SiSV (MONDAL et al., 1974). Whether reverse transcriptase activity related to BaEV was also present could not be determined at that time since the appropriate antiserum was unavailable.

In December 1974, fresh specimens of blood and marrow cells were drawn from patient HL23 and established in tissue culture. At this point the patient was beginning to relapse from chemotherapy-induced remission. Virus was detected in the cultured blood leukocytes by electron microscopy but not in the cultured marrow cells. Virus transmission attempts with the blood leukocytes were unsuccessful, but virus was transmitted from the bone marrow specimen.

In January 1975, new specimens of blood and marrow cells were drawn and established in tissue culture. The patient was in full relapse and on heavy chemotherapy. Virus was transmitted from the blood leukocytes but not from the bone marrow cells.

A portion of these uncultured blood leukocytes was reserved for biochemistry. Virus-like particles were detected in the cytoplasm of these leukocytes as were nucleic acids related to the genomes of both SiSV and BaEV (REITZ et al., 1976). These are likely to be two separate sets of nucleic acid sequences (CHAN et al., 1976).

In summary, virus was repeatedly detected in and transmitted from cultured blood and marrow leukocytes of patient HL23 but not from cells of many other patients. Moreover, the fresh, uncultured blood leukocytes of patient HL23 harbored SiSV- and/or BaEV-related markers in two separate blood samples collected more than one year apart.

When the transmitted virus was analyzed immunologically and by molecular hybridization, several groups noticed two viral components, one SiSV-like and another BaEV-like (CHAN et al., 1976; REITZ et al., 1976). Early studies could find no differences between HL23V components and their laboratory counterparts. Thus, reverse transcriptase, gp70, p12, and p30 and the RNA genomes of SiSV and BaEV were experimentally identical to the comparative components of HL23V. More recently, small differences have been noticed, especially in analyses of the sequence of "strong stop" cDNA made on the RNA of the viruses (W. Haseltine, personal communication). Nonetheless, the differences are small and not relatable to biological origin at present. The question of whether a virus isolate ostensibly from humans must necessarily be distinguishable from viruses obtained from nonhumans in order to qualify as a "human" virus will be dealt with later.

(4) *The Rijswijk Viruses.* NOOTER, BENTVELZEN and their colleagues have isolated several viruses ostensibly from marrow cells of leukemic children. The first, reported by NOOTER et al. (1975) came from a four year old boy with lymphosarcoma that progressed to lymphoblastic leukemia. Type-C particles were seen in media after cultivation of the child's marrow cells for three days in the presence of phytohemagglutinin. The leukemic cells were cocultivated with XC cells (rat cells transformed by but not producing Rous sarcoma virus) wherein syncytia were produced. No syncytia were produced after cocultivation with five other samples from normal or leukemic individuals. The virus produced by the XC cells contained determinants related to SiSV and MuLV$_R$ (Rauscher murine leukemia virus) and was transmitted to HEK human embryonic kidney cells. Only MuLV$_R$-related antigens were assayed in the virus from HEK cells, and they were present. It is not known whether two sets of determinants existed, one related to SiSV and another to MuLV$_R$, or whether one determinant reacted with both antisera. No antigens related to rat endogenous virus were detected. Antigens related to BaEV were not analyzed. The infected cells eventually stopped producing virus, making further analyses difficult.

Recently, NOOTER et al. (1977 and 1978) reported the isolation of viruses from two new cases of childhood leukemia. One (patient 21875) was a three year old child having tonsillar lymphoblastic leukemia with marrow infiltration. The second (patient 221075) was a five year old child having acute lymphoblastic leukemia with involvement of the bone marrow. Marrow cells were cocultivated with uninfected A7573 canine thymus cells, resulting in a transient burst of extracellular, particulate, reverse transcriptase activity at passages 3—5. The infected A7573 cells were taken at passage 5 and cocultivated separately with two transformed nonproducer cell lines, a human osteogenic sarcoma cell line transformed by Kirsten murine sarcoma virus (R-970-5) and a normal rat kidney line transformed by the same virus (KNRK). These cocultures exhibited extracellular, particulate reverse transcriptase activity after 8 days of culture and released transforming virus, capable to forming foci on several cell lines including SIRC rabbit corneal cells. These viruses are then available in two nonproducer cell lines (tertiary host cells) and several quaternary hosts. The secondary host cells, A 7573, no longer produce virus and the primary marrow cells never did produce infectious virus.

The antigenic properties of the virus from patient 21875, was tested and found to carry SiSV-related antigens. The test was performed on virus passed through KNRK to SIRC cells, virus called SAK-21 virus. BaEV-related antigens were not examined. Molecular hybridization experiments, carried out by R. G. Smith and M. S. Reitz (personal communication) confirm the presence of an SiSV-related component. BaEV-related nucleotide sequences have not yet been found.

It is the interpretation of the authors that the transforming potential of the SAK-21 virus derives from the Kirsten virus component of the nonproducer cells and that the replicating activity results from an SiSV-related helper virus from the leukemic patient. The procedure for obtaining virus from human leukemic cells described above is frequently successful, at least with childhood leukemia (P. Bentvelzen, personal communication).

(5) *The L-104 Virus.* GABELMAN *et al.* (1975) reported the isolation of a virus from a 66-year old male with a pulmonary adenocarcinoma and concurrent chronic lymphocytic leukemia. Primary lung tumor cells (L1 cells) were established in culture but showed no evidence of virus components. Cocultivation of L1 cells with XC cells (rat cells infected by but not producing Rous sarcoma virus) gave rise to stable dimorphic cultures consisting of epitheloid and fibroblast cell types (L104 cells) and producing type-C virus (L104 virus). The virus shares antigenic and nucleic acid homology with SiSV. No BaEV-related component has been reported.

(6) *The HCCL Virus.* This virus was isolated from a 60-year old man with metastatic adenocarcinoma of the stomach by BALABANOVA *et al.* (1975). Primary tumor cells were carried as a monolayer in culture. Most of the cultures appeared to be fibroblastic, spindle-shaped cells, but one bottle developed foci of polyploid, epitheloid cells (HCCL cells). This culture produced particles having a buoyant density in sucrose of 1.18 gm/ml which were then labeled with radioactive uridine. Particles of similar density contained a 70S RNA and a DNA-synthesizing activity (presumably reverse transcriptase) capable of using the endogenous 70S RNA as a template for cDNA synthesis. Particles released into the media by HCCL cells caused foci on human embryonic muscle but not on skin or kidney cells. The transformed muscle cells released virus, but the level of production is so low as to hamper detailed investigations (Y. Becker, personal communication). The virus carried antigens related to SiSV proteins (S. Spiegelman, personal communication).

(7) *The DHL-1 Virus.* This virus was isolated from a 10-year old male with diffuse histiocytic lymphoma (KAPLAN *et al.*, 1977). Cells from the pleural effusion of this patient were put into culture in 1974 (EPSTEIN and KAPLAN, 1974). Cytogenetic, immunological and histochemical markers defined the SU-DHL-1 line as a tumor line. Microsomal particles having a buoyant density of about 1.14 g/ml in sucrose contained a DNA polymerase capable of using oligo(dT)·poly(rA) as primer-template and an endogenous template-primer complex of high molecular

weight (KAPLAN *et al.*, 1977). Low levels of DNA polymerase were also excreted into the medium. The extracellular DNA polymerase, when purified, had properties of true reverse transcriptase and was related in amino acid sequence to reverse transcriptase of SiSV and BaEV, but clearly differed from all standard laboratory viruses tested (GOODENOW and KAPLAN, 1979). Thus, its presence in the SU-DHL-1 cell is not easily explained by contamination. Whether it is a viral-human recombinant enzyme is not yet konwn. To our knowledge it is not known whether the DHL-1 agent is capable of completing a life cycle nor has a genetic relationship been established between the DHL-1 genome, if it has one, and viral or cell genes. Nevertheless, the DHL-1 "virus" represents the only extant example which shows demonstrable differences from animal retroviruses.

(8) *The HEL-12 and HEL-1 viruses.* These and the HL23 viruses are among the best studied of viruses that still can be considered candidate human viruses. Unlike the viruses previously described, the HEL-12 and HEL-1 viruses were obtained from tissues of apparently normal (nonmalignant), human embryos. Primary cultures of HEL-12 cells were established from lung tissue of an 8 week-old female embryo (PANEM *et al.*, 1975). The mother, a 31-year old black, gravida IX, para VII, had no history of cancer. Primary cultures of HEL-1 cells were established in 1969 from the lungs of a sixteen week-old male embryo. The mother was a 34-year old black, gravida V, para V, who developed breast carcinoma in 1959. She was treated by radical mastectomy and received X-irradiation in 1959.

PROCHOWNIK, PANEM, KIRSTEN and their colleagues reported that both the HEL-12 and HEL-1 cells began to produce virus after about 35 and 45 passages in culture, respectively. In both cases the viruses shared antigenic relatedness with both SiSV and BaEV (PANEM *et al.*, 1977; PROCHOWNIK *et al.*, 1979). The presence of an SiSV-related component has been confirmed in the HEL-12 virus by molecular hybridization (unpublished results, M. S. Reitz and R. G. Smith). These viruses are infectious viruses capable of replicating in secondary cells and in the case of HEL-12 virus the SiSV- and BaEV-related components are not separable (S. Panem, personal communication).

The HEL-12 and HEL-1 cells are capable of supporting virus production for only about 10 passages. Some 5—10 passages earlier, intracellular viral antigens are spontaneously produced (PANEM *et al.*, 1975; PROCHOWNIK *et al.*, 1979) and production of virus particles can be induced with iododeoxyuridine, but there is little or no spontaneous production of infectious virus or virus particles. At an earlier time yet, though intracellular antigens are not produced spontaneously, they are synthesized following iododeoxyuridine treatment. Thus, the appearance of intracellular viral proteins or extracellular virus exhibits an inducible phase preceding a period of spontaneous production.

PROCHOWNIK *et al.*, emphasize that these transient events and their intricate scheduling may explain the failure of others to have made similar observations. Additionally, not every human embryonic strain tested yielded virus. Virus production by the HEL-12 and HEL-1 strains was reproducible, using several early freezes of cells and there is no reason to suspect virus contamination of these stocks, especially since the HEL-12 cells were frozen in 1969—1971 in a laboratory not working on RNA tumor viruses.

General. We support the view that most or all of the viruses isolated from human materials since 1975 do not reflect laboratory contaminants. Great care has been taken to eliminate this possibility, to the extent of constructing special virus-free laboratories for isolation of human agents. The fact that the viruses are always specifically related to primate retroviruses (simian sarcoma virus-gibbon ape leukemia virus) and that reverse transcriptase and nucleic acids related to the same specific retroviruses are detectable in fresh, uncultured human tumors supports this conclusion. This is especially true since the animal retroviruses are known to be passed among primates and are capable of causing leukemia and sarcoma in them. It is not surprising that infectious viruses are only rarely isolated, considering that even laboratory-reared retroviruses are generally defective and considering that in some cat leukemias feline leukemia virus cannot be detected, even though the virus is known to be the etiologic agent for the disease (ESSEX *et al.*, 1975). It may now be time to expect that retroviruses are etiologically related to human tumors and that the recombinant model of tumorigenesis developed from work on murine retroviruses will be applicable to man.

VIII. Retrovirus Integration, Growth Regulation and Human Cancer

A reasonable model for tumorigenesis, one which confers a common mode of action upon retroviruses, DNA-containing viruses, chemical carcinogens and other tumor-eliciting agents takes advantage of the essence of both the oncogene-virogene theory (HUEBNER and TODARO, 1969) and the protovirus hypothesis (TEMIN, 1971) and puts them in the context of our current state of knowledge.

It is certainly true that normal cell genomes contain genes that code for functions sufficient to create infectious retroviruses (see Section II on Origin of Retroviruses). It is probably true that part of the virogene genetic apparatus involves growth-regulating functions, for example, the envelope glycoprotein gene (*env*). The involvement of virogenes in oncogenesis can be pictured as direct if, for example, virogenes carry growth-regulating functions as part of their intrinsic properties, or if they can translocate and form hybrid genes consisting of a portion of the virogene fused with growth-regulating cell genes. The involvement can be pictured as indirect if virogenes affect the expression of growth-regulating genes or if they abort the utilization of products of growth-regulating genes.

The number and nature of growth-regulating genes that can be modified and then cause cancer has only recently become amenable to experimental attack. *Src* gene products are being synthesized *in vitro*, but a catalogue of their properties is not yet available. The role of the postulated growth-promoting genes in normal differentiation is also obscure, and, to our knowledge, there is no information available on this point. TODARO and his colleagues are documenting abnormalities in the recognition of low molecular weight growth factor proteins by transformed cells (SHERWIN, SLISKI, and TODARO, 1979), but whether these growth factors correspond to oncogene products is not known.

To maintain a "common cause" hypothesis for the origin of cancer, it is mandatory to invoke an alteration in the DNA of a cell since sarcoma-causing retroviruses carry a transforming gene which is an altered cell gene (see Section II and Section VI). Temin (1971) discussed virogene translocation and modification as a means of generating an oncogenic virus from a nononcogenic "protovirus". Cell gene modification by DNA-containing viruses and by chemical carcinogens can be accommodated by this model, if one assumes that the role of virogene translocation in carcinogenesis is to modify cell genes and that viruses are site-specific while chemicals are site-unspecific.

The only obvious model wherein the transformed cell is *not* the genetically altered cell, specifies that there exists a nontransformed cell that produces a diffusible growth-regulating substance. The cell it regulates appears malignant if the growth-regulating substance is altered. In this model the producer clonal cell undergoes the genetic change while the transformed cell is genetically normal. Transplantable tumors are difficult to fit into this model as are clonal cancers.

We favor a model wherein a virogene or a portion of it translocates to the environs of a growth-promoting gene, altering the amount or structure of its gene product thereby altering its function in particular settings; *i. e.* in a committed cell. We call this translocation a *microrearrangement*. The structure of the RNA genome of Rous sarcoma virus (Fig. 3) is evidence that this type of translocation occurs and can result in neoplasia. The genome of Rous sarcoma virus carries genes corresponding to chicken RAV_0 virogenes covalently linked to a *src* gene. The *src* gene is required for the transforming activity of the virus and represents a modified form of a normal chicken gene, termed *protosrc* (see Section II and Section IV). We picture Rous sarcoma virus as a recombinant between two types of cell genes: a virogene and a growth-regulating gene. Its origin can be viewed as a DNA microrearrangement; the translocation of a chicken virogene to the environs of a growth-regulating gene, *protosrc*.

This situation is not restricted to Rous sarcoma virus. Most oncogenic viruses are recombinants between viral genes and cell genes (see Section VI on Relatedness Among Retroviruses: Recombinant Viruses). In most cases it has not been rigorously proven that the cell genes are growth-regulating genes, but in the case of Moloney and Kirsten murine sarcoma viruses, the initial evidence favors this idea.

In some cases the recombination events occur between viral genes from one species and cell genes from another (see Sections V and VI on Horizontal Transmission of Retroviruses Among Animals and on Relatedness Among Retroviruses: Recombinant Viruses). Interspecies transfer of endogenous retroviruses appears to be common and indeed the transfer appears to enhance the probability that a virus will become tumorigenic.

It is most improbable that an agent like Rous sarcoma virus, a replicating virus carrying cancer-causing genes, is the etiology of all or most cancers. In Fig. 49, transformation by Rous sarcoma virus is case A and is pictured to be direct, often resulting in a virus-producing tumor. In Fig. 50, transformation by Rous sarcoma virus takes a normal cell (1) directly to a tumor cell carrying a malignant microrearrangement (5). More likely, a retrovirus which causes malignancies in outbred populations, will be less efficient, replicate poorly and only rarely cause cancer after infection. Imagine, for example, the consequence of infection of a chicken cell by RAV_0. In Fig. 49 transformation "caused" by RAV_0 is case B and is pictured to be indirect, requiring many genetic alternations in order to produce a tumor. Almost always, this infection has no discernible effect. However, we know that at least one time it recombined with cell genes to form an oncogenic virus, Rous sarcoma virus. We suspect also that other recombinant oncogenic viruses can occasionally be formed by infection of avian cells with RAV_0

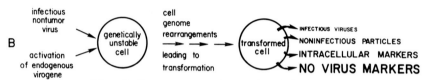

Fig. 49. Consequences of retrovirus infection.

Scheme A shows the results of infection by oncogenic retroviruses, often giving rise to transformed cells which produce infections virus. Transformation is a direct consequence of infection; the nature of the infection dictates whether virus genes are expressed or whether active viruses are produced. Scheme B shows the results of infection by nononcogenic retroviruses or of virogene activation. In this case several steps are required to produce cell transformation (see Fig. 50 and text) and infectious viruses are usually not produced by the cancer cell

NORMALCY	PRELEUKEMIA	LEUKEMIA
diploid cells with controlled growth	generation of proliferating clones	malignant clone expansion
① mutagenesis retrovirus infection DNA virus infection	③ MICROREARRANGEMENTS ② DNA mobilization ④ Karyotype alteration (macrorearrangements)-? ?	⑤ malignant microrearrangement ⑦ regulatory ⑧ growth change change ⑥ malignant macrorearrangement
FIRST HIT	SECOND HIT	THIRD HIT HIT CONSEQUENCES

(1)   repeated DNA serves as integration sites, cytogenetics normal.
(2) and (3) repeated DNA furnishes recombination sequence, cytogenetics normal.
(4)   repeated DNA furnish loci of homology, cytogenetic abnormalities appear and increase with time.
(5)   malignant microrearrangement involving growth-regulating gene; proliferating clone usually has (nonspecific) macrorearrangement.
(6)   malignant macrorearrangement involving growth-regulating gene; proliferating clone has specific cytogenetic abnormality.
(7) and (8) affected by biological response modifiers, e.g. hormones, interferon, growth factors.

Fig. 50. Hypothesis for carcinogenesis.

Infection of normal individuals by retroviruses or other viruses leads to genetic instability. Genetic instability characterizes the precancer state. New gene combinations are formed frequently during precancer. Some gene combinations alter growth patterns, leading to cancer. See text for fuller explanation (see Section VI on Relatedness Among Retroviruses: Recombinant Viruses). These oncogenic viruses are usually discovered by examining extracts of tumors for infectious virus. How much more often is a noninfectious recombinant formed ? Probably this is the most common result of recombination of viral genes with cell genes. See GILLESPIE, SAXINGER and GALLO (1975) for a discussion on this point.

It is likely that retrovirus-induced cancer, which occurs naturally in outbred animals, is usually caused by infection of an animal with a nontumorigenic virus which, by recombination with and modification of growth-promoting cell genes becomes tumorigenic after infection (Fig. 49, part B). The initial infection need not involve the target cell but could involve any cell of the body that promotes virus replication. The virus released from this first infection would promote many new infections in other cells of the body, each new infection resulting in recombination events for integration of the virus genome into the infected cell chromosome. Some integration may lead to DNA mobilization, i. e. to DNA rearrangements not requiring infectious virus (see later). If one of these secondary infections results in the rearrangement of a growth-promoting gene, transformation of that cell is probable.

The secondary infections were thought to be nonclassical, in the sense that only a portion of the virus genome need be inserted, a portion sufficient to modify the activity of the growth-regulating gene, but a portion so small as to be undetectable by current methods of molecular biology (GILLESPIE and GALLO, 1975). It is clear now that integration of small portions of infecting viral genomes is common (see Section IV on Organization of Infectious Retrovirus Genes); it remains to be seen whether these partial integrations can result in cancer.

It is thought that there are many integration sites in cell chromosomes for accepting the genomes of infecting RNA tumor viruses (*int* sites; see Introduction). In the context of the present model for tumorigenesis, we must invoke some degree of specificity to the integration event, otherwise retroviruses would be no more efficient at causing cancer by modifying growth-regulating genes than are mutagenic chemicals. The use of repeated DNA sequences as *int* sites satisfies the multiplicity-specificity problem. Sites containing repeated DNA sequences in mammalian genomes number in the thousands or tens of thousands, small numbers compared with the millions of potential genes in a mammalian genome. Should an infecting virus acquire a cellular repeated DNA sequence as part of its genome, then it acquires homology with specific yet multiple *int* sites (see Section V). When the virus acquires a cellular repeated DNA sequence that is also found in the vicinity of a growth-promoting gene, the virus becomes able to modify the growth-promoting gene with a reasonably high probability. Thus, among the microarrangements from secondary infections by such a virus, there will be some leading to transformation.

KLEIN (1979) recently reminded us that any model for carcinogenesis must take into account chromosomal abnormalities often seen in tumor cells. He proposed a three-hit model for human lymphomas involving an initial infection by Epstein-Barr virus followed later by a secondary chromosome rearrangement.

BRODSKY (1973) had earlier proposed that in leukemia the first hit, a retrovirus infection, would lead to a preleukemic state; the individual thereby incurring a high risk of developing acute leukemia. KLEIN (1979) reiterated this thought and postulated further that long-lived, preneoplastic, nontumorigenic cells develop additional genetic changes, including specific karyotypic alternations, which may change genetic control of cellular responses to growth-regulating forces. Repeated DNA could mediate both "hits"; it could serve as integration sites for viral infection and as loci of homology for chromosome exchanges. Virus infection could somehow perturb repeated DNA geometry such that abnormal chromosome pairing is encouraged and translocations frequently occur.

Figure 50 diagrams a specific hypothesis to this effect. The hypothesis derives from data pertaining primarily to leukemia-lymphoma but is probably a generalized theory. This model is a "three-hit" scheme. The first hit is usually virus infection which produces a viremia. Individuals at this stage are not preleukemic, they are normal but with a risk of progressing to preleukemia. Individuals who handle the infection poorly increase the chance of a secondary infection of a type that can eventually lead to malignancy. The type of secondary infection that can lead to malignancy is one where infecting virus genes integrate in repeated DNA in a way that causes mobilization of that DNA region.

DNA mobilization is the second hit. DNA mobilization occurs when the virus integrates next to a recombination-promoting sequence in repeated DNA and causes frequent microrearrangements through its "reintegration" at multiple chromosomal sites. This process can occur indefinitely without detectable effect or, more probably, will occasionally result in a specific microrearrangement of a growth-regulating gene, leading to cancer. This type of microrearrangement we call a malignant microrearrangement. Also occasionally, macrorearrangements occur by mobilization of loci consisting of tandemly-ordered repeated DNA and are detected as karyotypic alterations. Initially, these macrorearrangements are more or less random. Cells with a specific karyotype may accumulate when a malignant clone proliferates. Certain macrorearrangements (malignant macrorearrangements) contain specific microrearrangements within them and appear to be causally linked to cancer. In either case the malignant rearrangement leads to regulatory changes which lead to growth changes that characterize the phenotype of the individual cancers.

Radiation and carcinogens can act directly or indirectly at any step. They can activate endogenous virogenes (step 1), causing reintegration and indirectly causing microrearrangements. They may directly mobilize DNA. Or in the most extreme case a carcinogen can directly alter a growth-regulating gene, by causing a malignant microrearrangement.

It follows from the model that cancer therapy which is most likely to be of value will control the processes which lead to the malignant rearrangements in addition to controlling the proliferating cell clones (Fig. 50). Otherwise, continued rearrangements will occur during and after therapy and the probability of relapse will be high. The use of antiviral agents to control chronic viral infections may be useful in preventing certain cancers. For example, if chronic hepatitis B infection is a cause of hepatomas, then successful antiviral therapy with interferon should prevent the subsequent development of liver cancer. It is interesting to note that interferon indeed does seem to be an antitumor agent. It will be even more interesting to see whether its action includes control of a premalignant pool of cells, in addition to its "antitumor" or "antiproliferative" effects on malignant cells.

References

AARONSON, S., TODARO, G., SCOLNICK, E.: Induction of murine C-type viruses from clonal lines of virus-free BALB/3T3 cells. Science **174**, 157—159 (1971).

ADAMS, S. L., SOBEL, M. E., HOWARD, B. H., OLDEN, K., JAMADA, K. M., DE CROMBRUGGHE, B., PASTAN, I.: Levels of translatable mRNAs for cell surface protein, collagen precursors, and two membrane proteins are altered in Rous sarcoma virustransformed chicken embryo fibroblasts. Proc. nat. Acad. Sci. (Wash.) **74**, 3399—3403 (1977).

ALTANER, C., TEMIN, H. M.: Carcinogenesis by RNA sarcoma viruses XII. Virology **40**, 118—134 (1970).

ANDERSON, G. P., ROBBINS, K. C.: Rat sequences of the Kirsten and Harvey murine sarcoma virus genomes: Nature, origin, and expression in rat tumor RNA. J. Virol. **17**, 335—351 (1976).

AOKI, T., LIU, M., WALLING, M. J., BUSHAR, G. S., BRANDCHAFT, P. B., KAWAKAMI, T. G.: Specificity of naturally occurring antibody in normal gibbon serum. Science **191**, 1180—1183 (1976).

AOKI, T., WALLING, M. J., BUSHAR, G. S., LIU, M., HEU, K. C.: Natural antibodies in sera from healthy humans to antigens on surfaces of type C RNA viruses and cells from primates. Proc. nat. Acad. Sci. (Wash.) **73**, 2491—2495 (1976).

ARNHEIM, N., SOUTHERN, E. M.: Heterogeneity of the ribosomal genes of mice and men. Cell **11**, 363—370 (1977).

BADER, J. P., STECK, T. L.: Analysis of the ribonucleic acid of murine leukemia virus. J. Virol. **4**, 454—459 (1969).

BALABANOVA, H., KOTLER, M., BECKER, Y.: Transformation of cultured human embryonic fibroblasts by oncornavirus-like particles released from a human carcinoma cell line. Proc. nat. Acad. Sci. (Wash.) **72**, 2794—2798 (1975).

BALTIMORE, D.: RNA-dependent DNA polymerase in virions of RNA tumor viruses. Nature (Lond.) **226**, 1209—1211 (1970).

BALUDA, M. A.: Widespread presence in chickens of DNA complementary to the RNA genome of avian leukosis viruses. Proc. nat. Acad. Sci. (Wash.) **69**, 576—580 (1972).

BALUDA, M. A., NAYAK, D. P.: DNA complementary to viral RNA in leukemic cells induced by avian myeloblastosis virus. Proc. nat. Acad. Sci. (Wash.) **66**, 329—336 (1970).

BALUDA, M. A., ROY-BURMAN, P.: Partial characterization of RD114 by DNA-RNA hybridization studies. Nature (New Biol.) **244**, 59—62 (1973).

BARBACID, M., TRONICK, S. R., AARONSON, S. A.: Isolation and characterization of an endogenous type C RNA virus of mink cells. J. Virol. **25**, 129—137 (1978).

BASSIN, R. H., PHILIPS, L. A., KRAMER, M. J., HAAPALA, D. K., PEEBLES, P. T., NOMURA, S., FISCHINGER, P. J.: Transformation of mouse 3T3 cells by murine sarcoma virus: release of virus-like particles in the absence of replicating murine leukemia helper virus. Proc. nat. Acad. Sci. (Wash.) **68**, 1520—1524 (1974).

BASSIN, R. H., DURAN-TROISE, G., REIN, A., GERWIN, B.: Loss of Fv-1 restriction in BALB/3T3 cells following infection with a single N-tropic murine leukemia virus particle. J. Virol. **26**, 306—315 (1977).

BASSIN, R. H., TUTTLE, N., FISCHINGER, P. J.: Isolation of murine sarcoma virus-transformed mouse cells which are negative for leukemia virus from agar suspension cultures. Int. J. Cancer **6**, 95—107 (1970).

BATTULA, N., TEMIN, H. M.: Infectious DNA of spleen necrosis virus is integrated at a single site in the DNA of chronically infected chicken fibroblasts. Proc. nat. Acad. Sci., (Wash.) **74**, 281—285 (1977).

BAXT, W.: Sequences present in both human leukemic cell nuclear DNA and Rauscher leukemia virus. Proc. nat. Acad. Sci. (Wash.) **71**, 2853 (1974).

BAXT, W., HEHLMAN, R., SPIEGELMAN, S.: Human leukemic cells contain reverse transcriptase associated with a high molecular weight virus-related RNA. Nature (New Biol.) **240**, 72—75 (1972).

BAXT, W., SPIEGELMAN, S.: Nuclear DNA sequences present in human leukemic cells and absent in normal leukocytes. Proc. nat. Acad. Sci. (Wash.) **69**, 3737—3741 (1972).

BAXTER-GABBARD, K. L., CAMPBELL, W. F., PADGETT, F., RAITANO-FENTON, A., LEVINE, A. S.: Avian reticuloendotheliosis (strain T). II. Biochemical and biophysical properties. Avian Dis. **15**, 850—862 (1971).

BEEMON, K., HUNTER, T.: *In vitro* translation yields a possible Rous sarcoma virus *src* gene product. Proc. nat. Acad. Sci. (Wash.) **74**, 3302—3306 (1977).

BENDER, W., DAVIDSON, N.: Mapping of poly(A) sequences in the electron microscope reveals unusual structure on type C oncornavirus RNA molecules. Cell **7**, 595—607 (1976).

BENTVELZEN, P., DAAMS, J. H.: Hereditary infection with mammary tumor viruses in mice. J. nat. Cancer Inst. **43**, 1025—1035 (1969).

BENTVELZEN, P., DAAMS, J. H., HAGEMAN, P., CALAFAT, J.: Genetic transmission of viruses that incite mammary tumors in mice. Proc. nat. Acad. Sci. (Wash.) **67**, 377—384 (1970).

BENVENISTE, R. E., LIEBER, M. M., LIVINGSTON, D. M., SHERR, C. J., TODARO, G. J., KALTER, S. S.: Infectious C type virus isolated from baboon placenta. Nature (Lond.) **248**, 17—20 (1974).

BENVENISTE, R. E., SCOLNICK, E. M.: RNA in mammalian sarcoma virus transformed nonproducer cells homologous to murine leukemia virus RNA. Virology **51**, 370 to 382 (1973).

BENVENISTE, R. E., TODARO, G. J.: Homology between type-C viruses of various species as determined by molecular hybridization. Proc. nat. Acad. Sci. (Wash.) **70**, 3316—3320 (1973).

BENVENISTE, R. E., TODARO, G. J.: Evolution of type-C viral genes: inheritance of exogenously acquired viral genes. Nature (Lond.) **252**, 456—459 (1974).

BENVENISTE, R. E., TODARO, G. J.: Multiple divergent copies of endogenous C-type virogenes in mammalian cells. Nature (Lond.) **252**, 170—172 (1974).

BENVENISTE, R. E., TODARO, G. J.: Evolution of type-C viral genes I. Nucleic acid from baboon type-C virus as a measure of divergence among primate species. Proc. nat. Acad. Sci. (Wash.) **71**, 4513—4518 (1974).

BENVENISTE, R. E., TODARO, G. J.: Evolution of type-C viral genes: Origin of feline leukemia virus. Science **190**, 886—888 (1975).

BENVENISTE, R. E., TODARO, G. J.: Evolution of type-C viral genes: Preservation of ancestral murine type-C viral sequences in pig cellular DNA. Proc. nat. Acad. Sci. (Wash.) **72**, 4090—4094 (1975).

BENVENISTE, R. E., TODARO, G. J.: Segregation of RD114 and FeLV-related sequences in crosses between domestic cat and leopard cat. Nature (Lond.) **257**, 506—508 (1975).

BENVENISTE, R. E., TODARO, G. J.: Evolution of type-C viral genes: evidence for an Asian origin of man. Nature (Lond.) **261**, 101—108 (1976).

BENVENISTE, R. E., SHERR, C., TODARO, G. J.: Evolution of type-C viral genes: origin of feline leukemia virus. Science **190**, 886—888 (1975).

BENVENISTE, R. E., HEINEMANN, R., WILSON, G. L., CALLAHAN, R., TODARO, G. J.: Detection of baboon type-C viral sequences in various primate tissues by molecular hybridization. J. Virol. **14**, 56—67 (1974).

BENVENISTE, R. E., CALLAHAN, R., SHERR, C. J., CHAPMAN, V., TODARO, G. J.: Two
distinct endogenous type-C viruses isolated from the Asian rodent *Mus cervicolor:*
Conservation of virogene sequences in related rodent species. J. Virol. **21**, 849—862
(1977).
BISHOP, J. M.: Retroviruses. Ann. Rev. Biochem. **47**, 35—88 (1978).
BISTER, K., DUESBERG, P. H.: Structure and specific sequences of avian erythroblastosis virus RNA: Evidence for multiple classes of transforming genes among avian
tumor viruses. Proc. nat. Acad. Sci. (Wash.) **76**, 5023—5027 (1979).
BONDURANT, M., RAMABHADRAN, R., GREEN, M., WOLD, M.: 'Sarc' sequence tran-
scription in Moloney sarcoma virus-transformed nonproducer cell lines. J. Virol.
29, 76—82 (1979).
BRACK, C., HIRAMA, M., LENHARD-SCHULLER, R., TONEGAWA, S.: A complete
immunoglobin gene is created by somatic recombination. Cell **15**, 1—14 (1978).
BRACK, C., TONEGAWA, S.: Variable and constant parts of the immunoglobulin light
chain gene of a mouse myeloma cell are 1250 nontranslated bases apart. Proc. nat.
Acad. Sci. (Wash.) **74**, 5652—5656 (1977).
BREATHNACH, R., MANDEL, J. L., CHAMBON, P.: Ovalbumin gene is split in chicken
DNA. Nature (Lond.) **270**, 314—319 (1977).
BRITTEN, R. J., KOHNE, D. E.: Repeated sequences in DNA. Science **161**, 529—540
(1968).
BRODSKY, I.: Role of the megakaryocyte and platelet in the leukemic process in mice
and men — a review and hypothesis. J. nat. Cancer Inst. **51**, 329—335 (1973).
BRODY, R. S., McDONOUGH, S., FRYE, F. L., HARDY, W. D.: Epidemiology of feline
leukemia. In: Comparative Leukemia Research (DUTCHER, R. M., ed.), 333—342.
Basel: Karger 1970.
CALLAHAN, R., BENVENISTE, R. E., LIEBER, M. M., TODARO, G. J.: Nucleic acid
homology of murine type-C viral genes. J. Virol. **14**, 1394—1403 (1974).
CALLAHAN, R., LIEBER, M. M., TODARO, G. J., GRAVES, D. C., FERRER, J. F.: Bovine
leukemia virus genes in the DNA of leukemic cattle. Science **192**, 1005—1007
(1976).
CARTER, C. W., KRAUT, J.: A proposed model for interaction of polypeptides with
RNA. Proc. nat. Acad. Sci. (Wash.) **71**, 283—287 (1974).
CHAN, E., PETERS, W. P., SWEET, R. W., OHNO, T., KUFE, D. W., SPIEGELMAN, S.,
GALLO, R. C., GALLAGHER, R. E.: Characterization of a virus (HL23V) isolated
from cultured acute myelogenous leukemic cells. Nature (Lond.) **260**, 266—268
(1976).
CHATTOPADHYAY, S., HARTLEY, S. W., LANDER, M. R., KRAMER, B. S., ROWE, W. P.:
Biochemial characterization of the amphotropic group of murine leukemia viruses.
J. Virol. **26**, 29—39 (1978).
CHATTOPADHYAY, S. K., LOWY, D. S., TEICH, N. M., LEVINE, A. S., ROWE, W. P.:
Evidence that the AKR murine-leukemia-virus genome is complete in DNA of the
high-virus AKR mouse and incomplete in the DNA of "virus-negative" NIH
mouse. Proc. nat. Acad. Sci. (Wash.) **71**, 167—171 (1974).
CHATTOPADHYAY, S. K., LOWY, D. R., TEICH, N. M., LEVINE, A. S., ROWE. W. P.:
Qualitative and quantitative studies of AKR-type murine leukemia virus se-
quences in mouse DNA. Cold Spr. Harb. Symp. quant. Biol. **39**, 1085—1101 (1974).
CHURCH, G. M., SUSSMAN, J. L., KIM, S.: Secondary structural complementarity
between DNA and proteins. Proc. nat. Acad. Sci. (Wash.) **74**, 1458—1462 (1977).
CLAYMAN, C. H., MOSHARRAFA, E. T., ANDERSON, D. L., FARAS, A. J.: Circular forms
of DNA synthesized by Rous sarcoma virus *in vitro.* Science 206, 582—584 (1979).
COFFIN, J. M., BILLETER, M. A.: A physical map of the Rous sarcoma virus genome.
J. molec. Biol. **100**, 293—318 (1976).
COGGESHAL, L.: *Plasmodium lophurae*, a new species of malaria pathogenic for the
domestic fowl. Amer. J. Hyg. **27**, 615—618 (1938).
COLLET, M. S., ERIKSON, R. L.: Protein kinase activity associated with the avian
sarcoma virus src gene product. Proc. nat. Acad. Sci. (Wash.) **75**, 2021—2024
(1978).

COLLETT, M. S., ERIKSON, E., PURCHIO, A. F., BRUGGE, J. S., ERIKSON, R. L.: A normal cell protein similar in structure and function to the avian sarcoma virus transforming gene product. Proc. nat. Acad. Sci (Wash.) **76**, 3159—3163 (1979).

COLLET, M. S., KIERAS, R. M., FARAS, A. J.: Studies on the replication of reticuloendotheliosis virus: detection of viral-specific DNA sequences in infected chick cells. Virology **65**, 436—445 (1975).

CONSTANTINE, F. D., SCHELLER, R. H., BRITTEN, R. J., DAVIDSON, E. H.: Repetitive sequence transcripts in the mature sea urchin oocyte. Cell **15**, 173—187 (1978).

COOK, M. K.: Cultivation of a filterable agent associated with Marek's disease. J. nat. Cancer Inst. **43**, 203—212 (1969).

CREMER, N. E., TAYLOR, D. O. N., OHIRO, L. S., TEITZ, Y.: Transformation and virus production in normal rat thymus cells and those infected with Moloney leukemia virus. J. nat. Cancer Inst. **45**, 37—48 (1970).

DEARBORNE, E. H.: Filterable agents lethal for ducks. Proc. Soc. exp. Biol. (N. Y.) **63**, 48—49 (1946).

DePAULI, A., JOHNSEN, D. O., NOLL, W. W.: Granulocytic leukemia in white-handed gibbons. J. amer. Vet.-Med. Ass. **163**, 624—628 (1973).

DEWEY, M. J., MARTIN, D. W., MARTIN, G. R., MINTZ, B.: Mosaic mice with teratocarcinoma-derived mutant cells deficient in hypoxanthine phosphoribosyltransferase. Proc. nat. Acad. Sci. (Wash.) **74**, 5564—5568 (1977).

DINA, D., BEEMON, K.: Relationship between Moloney murine leukemia and sarcoma virus RNAs. J. Virol. **23**, 524—532 (1977).

DMOCHOWSKI, L., BOWEN, J. M.: Viruses and human cancer: The history and current status of ESP-1. Progr. exp. Tumor Res. **22**, 160—195 (1978).

DONEHOWER, L., WONG-STAAL, F., GILLESPIE, D.: Divergence of baboon endogenous type-C virogenes in primates using genomic viral RNA in molecular hybridization experiments. J. Virol. **21**, 932—941 (1977).

DOUGHERTY, R. M., DiSTEFANO, H. S.: Lack of relationship between infection with avian leukosis virus and presence of COFAL antigen in chick embryos. Virology **29**, 586—595 (1966).

DROHAN, W. N., SHOYAB, M., WALL, R., BALUDA, M. A.: Interspersion of sequences in avian myeloblastosis RNA that rapidly hybridizes with leukemic chicken cell DNA. J. Virol. **15**, 550—555 (1975).

DROHAN, W., COLCHER, D., SCHOCHETMAN, G., SCHLOM J.: Distribution of Mason Pfizer-specific sequences in the DNA of primates. J. Virol. **23** 36—43 (1977).

DUC-NGUYEN, J., ROSENBLUM, E. M., ZEIGEL, R. F.: Persistant infection of a rat kidney cell line with Rauscher murine leukemia virus. J. Bact. **92**, 1133—1140 (1966).

DUESBERG, P. H., ROBINSON, W. S.: Nucleic acid and proteins isolated from the Rauscher mouse leukemia virus (MLV). Proc. nat. Acad. Sci. (Wash.) **55**, 219—227 (1966).

DUESBERG, P. H., VOGT, P. K.: Avian acute leukemia viruses MC29 and MH2 share specific RNA sequences: Evidence for a second class of transforming genes. Proc. nat. Acad. Sci. (Wash.) **76**, 1633—1637 (1979).

DUNN, C. Y., AARONSON, S. A., STEPHENSON, J. R.: Interactions of chemical inducers and steroid enhancers of endogenous mouse type-C RNA viruses. Virology **66**, 579—588 (1975).

EAST, J. L., KNESEK, J. E., CHAN, J. C., DMOCHOWSKI, L.: Quantitative nucleotide sequence relationship of mammalian RNA tumor viruses. J. Virol. **15**, 1396—1408 (1975).

ELDER, J. H., GAUTSCH, J. W., JENSEN, F. C., LERNER, R. A., HARTLEY, J. W., ROWE, W. P.: Biochemical evidence that MCF murine leukemia viruses are envelope (env) gene recombinants. Proc. nat. Acad. Sci. (Wash.) **74**, 4676—4680 (1977).

EPSTEIN, A. L., KAPLAN, H. S.: Biology of the human malignant lymphomas. Cancer **34**, 1851—1872 (1974).

ESSEX, M.: Horizontally and vertically transmitted oncornaviruses of cats. Advanc. Cancer Res. **21**, 175—248 (1975).

Essex, M., Cotter, S. M., Hardy, W. D., Hess, P., Jarrett, W., Jarret, O., Mackey, L., Laird, H., Perryman, L., Olsen, R. G., Yohn, D. S.: Feline oncorna-virus-associated cell membrane antigen: IV. Antibody titers in cats with naturally-occurring leukemia, lymphoma and other diseases. J. nat. Cancer Inst. **55**, 463—467 (1975).

Essex, M., Klein, G., Snyder, S. P., Harrold, J. B.: Feline sarcoma virus induced tumors: correlation between humoral antibody and tumor regression. Nature (Lond.) **233**, 195—199 (1971).

Faller, D. V., Hopkins, N.: RNase Tl-resistant oligonucleotides of B-tropic murine leukemia viruses of BALB/c and five of its NB-tropic derivatives. J. Virol. **23**, 188—195 (1977).

Faller, D. V., Hopkins, N.: RNase Tl-resistant oligonucleotides of an N- and a B-tropic murine leukemia virus of BALB/c: evidence for recombination between these viruses. J. Virol. **24**, 609—617 (1977).

Faller, D. V., Hopkins, N.: Tl-oligonucleotide maps of N-, B- and B→NB-tropic murine leukemia virus derived from BALB/c. J. Virol. **26**, 143—152 (1978).

Frankel, A. E., Fischinger, P. J.: Rate of divergence of cellular sequences homo-logous to segments of Moloney sarcoma virus. J. Virol. **21**, 153—160 (1977).

Frankel, A., Haapala, D. K., Neubauer, R. L., Fischinger, P. J.: Elimination of the sarcoma genome from murine sarcoma virus transformed cat cells. Science **191**, 1264—1266 (1976).

Friend, C.: Cell-free transmission in adult Swiss mice of a disease having the cha-racter of a leukemia. J. exp. Med. **105**, 307—318 (1957).

Frisby, D. P., Weiss, R. A., Roussel, M., Stehelin, D.: The distribution of endo-genous retrovirus sequences in the DNA of galliform birds does not coincide with avian phylogenetic relationships. Cell **17**, 623—634 (1979).

Fry, K., Salser, W.: Nucleotide sequences of HS-α sattelite DNA from kangaroo rat *Dipodomys ordii* and characterization of similar sequences in other rodents. Cell **12**, 1069—1084 (1977).

Gabelman, N., Waxman, S., Smith, W., Douglas, S. D.: Appearance of C-type virus-like particles after co-cultivation of a human tumor cell line with rat (XC) cells. Int. J. Cancer **16**, 355—369 (1975).

Gallagher, R. E., Gallo, R. C.: Type-C RNA tumor virus isolated from cultured human acute myelogenous leukemia cells. Science **187**, 350 (1975).

Gallagher, R. E., Gallo, R. C.: Continuous production of complete type-C virus by exponentially growing cultured leukocytes from one of sixteen patients with myelogenous leukemia. In: Proceedings of the IInd International Congress on Pathological Physiology, Prague, Czechoslovakia, in press.

Gallagher, R. E., Salahuddin, S. Z., Hall, W. T., McCredie, K. B., Gallo, R. C.: Growth and differentiation in culture of leukemic leukocytes from a patient with acute myelogenous leukemia and reidentification of a type C virus. Proc. nat. Acad. Sci. (Wash.) **72**, 4137—4141 (1975).

Gallo, R. C., Gallagher, R. E., Wong-Staal, F., Aoki, T., Markham, P., Schet-ters, H., Ruscetti, F., Valerio, M., Saxinger, W. C., Smith, R. G., Gillespie, D., Reitz, M. S.: Biochemical and biological studies on a gibbon ape (*Hylobates* lar) with lymphocytic leukemia. Virology **84**, 359—373 (1978).

Gallo, R. C., Gallagher, R. E., Miller, N. R., Mondal, H., Saxinger, W. C., Mayer, R. J., Smith, R. G., Gillespie, D. H.: Relationships between components in primate RNA tumor viruses and in the cytoplasm of human leukemic cells: Implications to leukemogenesis. Cold Spr. Harb. Symp. quant. Biol. **34**, 933—961 (1974).

Gallo, R. C., Miller, N. R., Saxinger, W. C., Gillespie, D.: Primate RNA tumor virus-like DNA synthesized endogenously by RNA-dependent DNA polymerase in virus-like particles from fresh human acute leukemic blood cells. Proc. nat. Acad. Sci. (Wash.) **70**, 3219—3223 (1973).

Gallo, R. C., Yang, S. S., Ting, R. C.: RNA-dependent DNA polymerase of human acute leukemic cells. Nature (Lond.) **228**, 927—929 (1970).

GELB, L. D., AARONSON, S. A., MARTIN, M. A.: Heterogeneity of murine leukemia virus *in vitro* DNA; detection of viral DNA in mammalian cells. Science **172**, 1353—1355 (1971).

GETZ, M. J., REIMAN, H. M., SIEGAL, G. P., QUINLAN, J. J., PROPER, J., ELDER, P. K., MOSES, H. L.: Gene expression in chemically transformed mouse embryo cells: selective enhancement of the expression of type C RNA tumor virus genes. Cell **11**, 909—921 (1977).

GIANNI, A. M., SMOTKIN, D., WEINBERG, R.: Murine leukemia virus: Detection of unintegrated double-stranded forms of the provirus. Proc. nat. Acad. Sci. (Wash.) **77**, 447—451 (1975).

GILBOA, E., MITRA, S. W., GOFF, S., BALTIMORE, D.: A detailed model of reverse transcription and tests of crucial aspects. Cell **18**, 93—100 (1979).

GILDEN, R. V., OROSZLAN, S.: Structural and immunological relationship among mammalian type-C viruses. J. Amer. vet. med. Ass. **158**, 1099—1104 (1971).

GILLESPIE, D., GALLO, R. C.: RNA processing and the origin and evolution of RNA tumor viruses. Science **188**, 802—811 (1975).

GILLESPIE, D., GALLO, R. C.: New concepts of human myelogenous leukemia based on studies of the simian RNA tumor virus family. Bibl. haemat. (Basel) **43** (comparative Leukemia Research meeting 1975; CLEMMESEN, J., YOHN, D. S., eds.), 576—581. Basel: Karger 1976.

GILLESPIE, D., SAXINGER, W. C., GALLO, R. C.: Information transfer in cells infected by RNA tumor viruses and extension to human neoplasia. Progr. nucl. Acid Res. a. molec. Biol. **15**, 1—108 (1975).

GILLESPIE, D., GALLAGHER, R. E., SMITH, R. G., SAXINGER, W. C., GALLO, R. C.: On the evidence for type-C RNA tumor viruses information and virus-related reverse transcriptase in animals and human leukemic cells. In: Fundamental Aspects of Neoplasia (GOTTLIEB, A. A., PLESCIA, O. J., BISHOP, D. H. L., eds.), 3—27. New York: Springer 1975.

GILLESPIE, D., GILLESPIE, S., GALLO, R. C., EAST, J. L., DMOCHOWSKI, L.: Genetic origin of RD 114 and other RNA tumor viruses assayed by molecular hybridization. Nature (New Biol.) **244**, 51—54 (1973).

GILLESPIE, D., GILLESPIE, S., WONG-STAAL, F.: RNA-DNA hybridization applied to cancer research: Special-reference to RNA tumor viruses. Meth. Cancer Res. **11**, 205—245 (1975).

GILLESPIE, D., MARSHALL, S., GALLO, R. C.: RNA of RNA tumor viruses contains poly (A). Nature (New Biol.) **236**, 227—231 (1972).

GOLDBERG, R. J., LEVIN, R., PARKS, W. P., SCOLNICK, E. M.: Quantitative analysis of the rescue of RNA sequences by type-C ciruses. J. Virol. **17**, 43—50 (1976).

GOODENOW, R. S., KAPLAN, H. S.: Characterization of the reverse transcriptase of a type-C RNA virus produced by a human lymphoma cell line. Proc. nat. Acad. Sci. (Wash.) **76**, 4971—4975 (1979).

GROSS, L.: Spontaneous leukemia developing in C3H mice following inoculation, in infancy, with AK-leukemic extracts, or AK-embryos. Proc. Soc. exp. Biol. (N. Y.) **76**, 27—32 (1951).

GROSS, L.: Pathogenic properties and 'vertical' transmission of the mouse leukemia agent. Proc. Soc. exp. Biol. (N. Y.) **78**, 342—348 (1951).

GROSS, L.: The Rauscher virus: a mixture of the Friend virus and the mouse leukemia virus (Gross)? Acta haemat. (Basel) **35**, 200—213 (1966).

GROUDINE, M., WEINTRAUB, H.: Rous sarcoma virus activates embryonic globin genes in chicken fibroblasts. Proc. nat. Acad. Sci. (Wash.) **72**, 4464—4468 (1975).

GUNTAKA, R. V., MAHY, B. W. J., BISHOP, J. M., VARMUS, H. E.: Ethidium bromide inhibits appearance of closed circular viral DNA and integration of virus-specific DNA in duck cells infected by avian sarcoma virus. Nature (Lond.) **253**, 507—511 (1975).

HAAPALA, D. K., FISCHINGER, P. J.: Molecular relatedness of mammalian RNA tumor viruses as determined by DNA · RNA hybridization. Science **180**, 972—974 (1973).

HANAFUSA, T., HANAFUSA, H.: Isolation of leukosis-type virus from pheasant embryo cells: Possible presence of viral genes in cells. Virology 51, 247—251 (1973).

HANAFUSA, T., HANAFUSA, H., MIYAMOTO, T.: Recovery of a new virus from apparently normal chick cells by infection with avian tumor viruses. Proc. nat. Acad. Sci. (Wash.) 67, 1797—1803 (1970).

HARDY, W. D., OLD, L. J., HESS, P. W. ESSEX, M., COTTER, S. M.: Horizontal transmission of feline leukemia virus. Nature (Lond.) 244, 266—269 (1973).

HARTLEY, J. W., ROWE, W. P.: Naturally occurring murine leukemia viruses in wild mice: characterization of a new 'amphotropic' class. J. Virol. 19, 19—25 (1976).

HARTLEY, J. W., ROWE, W. P., HUEBNER, R. J.: Host range restrictions of murine leukemia viruses in mouse embryo cell cultures. J. Virol. 5, 221—225 (1970).

HARTLEY, J. W., WOLFORD, N. K., OLD, L. J., ROWE, W. P.: A new class of murine leukemia viruses associated with development of spontaneous lymphomas. Proc. nat. Acad. Sci. (Wash.) 74, 789—792 (1977).

HARVEY, J. J.: An unidentified virus which causes rapid production of tumors in mice. Nature (Lond.) 204, 1104—1105 (1964).

HASELTINE, W. A., MAXAM, A. M., GILBERT, W.: Rous sarcoma virus genome is terminally redundant: the 5' sequence. Proc. nat. Acad. Sci. (Wash.) 74, 989—993 (1977).

HAYWARD, W. S., HANAFUSA, H.: Detection of avian tumor virus RNA in uninfected chicken embryo cells. J. Virol. 11, 157—167 (1973).

HAYWARD, W. S., HANAFUSA, H.: Recombination between endogenous and exogenous RNA tumor virus genes as analyzed by nucleic acid hybridization. J. Virol. 122, 1367—1377 (1975).

HEHLMAN, R., KUFE, D. SPIEGELMAN, S.: RNA in human leukemic cells related to the RNA of a mouse leukemia virus. Proc. nat. Acad. Sci. (Wash.) 69, 435—439 (1972).

HILL, M., HILLOVA, J.: Virus recovery in chicken cells tested with Rous sarcoma cell DNA. Nature (New Biol.) 237, 35—39 (1972).

HOWK, R. S., TROXLER, D. H., LOWY, D., DUESBERG, P. H., SCOLNICK, E. M.: Identification of a 30S RNA with properties of a defective type-C virus in murine cells. J. Virol. 25, 115—123 (1978).

HUEBNER, R., TODARO, G. J.: Oncogenes of RNA tumor viruses as determinants of cancer. Proc. nat. Acad. Sci. (Wash.) 64, 1087—1091 (1969).

HUGHES, S. H., SHANK, P. R., SPECTOR, D. H., KUNG, H. J., BISHOP, J. M., VARMUS, H. E., VOGT, P. K., BREITMAN, M. L.: Proviruses of avian sarcoma virus are terminally redundant, co-extensive with unintegrated linear DNA and integrated at many sites. Cell 15, 1397—1410 (1978).

HUMPRIES, E., COFFIN, J. M.: Rate of virus-specific RNA synthesis in synchronized chicken embryo fibroblasts infected with avian leukosis viruses. J. Virol. 17, 393—401 (1976).

ILLMENSEE, K., MINTZ, B.: Totipotency and normal differentiation of single teratocarcinoma cells cloned by injection into blastocysts. Proc. nat. Acad. Sci. (Wash.) 73, 549—553 (1976).

JACOB, F., MONOD, J.: Genetic regulatory mechanisms in the synthesis of proteins. J. molec. Biol. 3, 318—356 (1961).

JACQUEMIN, P. C., SAXINGER, W. C., GALLO, R. C.: Surface antibodies of human myelogenous leukaemia leukocytes reactive with specific type-C viral reverse transcriptases. Nature (Lond.) 276, 230—234 (1978).

JAENISH, R.: Germline integration and Mendelian transmission of the exogenous Moloney leukemia virus. Proc. nat. Acad. Sci. (Wash.) 73, 1260—1264 (1976).

JAENISH, R.: Germline integration of Moloney leukemia virus: Effect of homozygosity at the M-MuLV locus. Cell 12, 691—696 (1977).

JAMJOOM, G. A., NASO, R. B., ARLINGHAUS, R. B.: Further characterization of intracellular polyprotein precursors of Rauscher leukemia virus. Virology 78, 11—34 (1977).

JARRETT, W. H. F., JARRETT, O., MACKEY, L., LAIRD, H., HARDY, W. P., ESSEX, M.: Horizontal transmission of leukemia virus and leukemia in the cat. J. nat. Cancer Inst. 51, 833—841 (1973).

JEFFREYS, A. J., FLAVELL, R. A.: The rabbit β-globin gene contains a large insert in the coding sequence. Cell **12**, 1097—1108 (1977).

JENSIK, S., HOEKSTRA, J., SILVER, S., NORTHROP, R. L., DEINHARDT, F.: The 60 to 70S RNA and reverse transcriptase of simian sarcoma and simian sarcoma-associated viruses. Intervirology **1**, 229—241 (1973).

JOLICOEUR, P., BALTIMORE, D.: Effect of Fv-1 gene product on proviral DNA formation and integration in cells infected with murine leukemia viruses. Proc. nat. Acad. Sci. (Wash.) **73**, 2236—2240 (1976).

JUNGHANS, R. P., HU, S., KNIGHT, C. A., DAVIDSON, N.: Heteroduplex analysis of avian RNA tumor viruses. Proc. nat. Acad. Sci. (Wash.) **74**, 477—481 (1977).

KALTER, S. S., HELMKE, R. J., HEBERLING, R. L., PANIGEL, M., FOWLER, A. K., STRICKLAND, J. E., HELLMAN, A.: C-type particles in normal human placentas. J. nat. Cancer Inst. **50**, 1081—1094 (1973).

KALTER, S. S., HELMKE, R. J., PANIGEL, M., HEBERLING, R. L., FELSBURG, P. J., AXELROD, L. R.: Observations on apparent C-type particles in baboon *(Papio cynocephalus)* placentas. Science **179**, 1332—1333 (1973).

KANG, C.-Y., TEMIN, H. M.: Lack of sequence homology among RNAs of avian leukosis-sarcoma viruses, reticuloendotheliosis viruses and chicken RNA-directed DNA polymerase activity. J. Virol. **12**, 1314—1324 (1973).

KANG, C.-Y., TEMIN, H. M.: Reticuloendotheliosis virus nucleic acid sequences in cellular DNA. J. Virol. **14**, 1179—1188 (1974).

KAPLAN, H. S., GOODENOW, R. S., EPSTEIN, A. L., GARTNER, S., DECLÈVE, A., ROSENTHAL, P. N.: Isolation of a C-type RNA virus from an established human histiocytic lymphoma cell line. Proc. nat. Acad. Sci. (Wash.) **74**, 2564—2568 (1977).

KAWAKAMI, T. G., BUCKLEY, P. M.: Antigenic studies in gibbon type-C viruses. Transplantation Proc. **6**, 193—196 (1974).

KAWAKAMI, T. G., BUCKLEY, P. M., McDOWELL, T. S., DePAOLI, A.: Antibodies to simian C-type virus antigen in sera of gibbons *(Hylobates* sp*)*. Nature (New Biol.) **246**, 105—107 (1973).

KAWAKAMI, T., HUFF, S., BUCKLEY, P., DUNGWORTH, S., SNYDER, S., GILDEN, R.: C type virus associated with gibbon lymphosarcoma. Nature (New Biol.) **235**, 170—171 (1972).

KAWASHIMA, K., IKEDA, H., HARTLEY, J. W., STOCKERT, E., ROWE, W. P., OLD, L. J.: Changes in expression of murine leukemia virus antigens and production of xenotropic virus in the late preleukemic period in AKR mice. Proc. nat. Acad. Sci. (Wash.) **73**, 4680—4684 (1976).

KEDES, L. H., COHN, R. H., LOWRY, J. C., CHEING, A. C. Y., COHEN, S. A.: The organization of sea urchin histone genes. Cell **6**, 359—369 (1975).

KESHET, E., TEMIN, H. M.: Nucleotide sequences derived from pheasant DNA in the genome of recombinant avian leukosis viruses with subgroup F specificity. J. Virol. **24**, 505—513 (1977).

KETTMAN, R., PORTETELLA, D., MAMMERICK, M., CLEUTER, Y., DECKEGEL, D., GODOUX, M., GHYSDAEL, J., BURNY, A., CHANTRENNE, H.: Bovine leukemia virus: an exogenous RNA oncogenic virus. Proc. nat. Acad. Sci. (Wash.) **73**, 1014—1018 (1976).

KEYL, H.-G.: A demonstrable local and geometric increase in the chromosomal DNA of Chironomus. Experientia (Basel) **21**, 191—193 (1965).

KIRSTEN, W. H., MAYER, L. A.: Morphological response to a murine erythroblastosis virus. J. nat. Cancer Inst. **39**, 311—319 (1967).

KIRSTEN, W. H., MAYER, L. A., WOLLMAN, R. L., PIERCE, M. I.: Studies on a murine erythroblastosis virus. J. nat. Cancer Inst. **38**, 117—139 (1967).

KLEIN, G.: Lymphoma development in mice and humans: Diversity of initiation is followed by convergent cytogenetic evolution. Proc. nat. Acad. Sci. (Wash.) **76**, 2442—2446 (1979)

KOPCHICK, J. J., JAMJOOM, G. A., WATSON, K. F., ARLINGHOUSE, R. B.: Biosynthesis of reverse transcriptase from Rauscher murine leukemia virus by synthesis and cleavage of a gag-pol read-through viral precursor polyprotein. Proc. nat. Acad. Sci. (Wash.) **75**, 2016—2020 (1978).

Koshy, R., Wong-Staal, F., Gallo, R. C., Hardy, W., Essex, M.: Distribution of feline leukemia virus DNA sequences in tissues of normal and leukemic domestic cats. Submitted to Virology.

Krzyek, R. A., Lau, A. F., Vogt, P. K., Faras, A. J.: Quantitation and localization of Rous sarcoma virus-specific RNA in transformed and revertant field vole cells. J. Virol. **25**, 518—526 (1978).

Kufe, K., Hehlmann, R., Spegelman, S.: Human sarcomas contain RNA related to the RNA of a mouse leukemia virus. Science **175**, 182—185 (1972).

Kung, H. J., Bailey, J. M., Davidson, N., Vogt, P. K., Nicolson, M. O., McAllister, R. M.: Electron microscope studies of tumor virus RNA. Cold Spr. Harb. Symp. quant. Biol. **39**, 827—834 (1974).

Kung, H. J., Hu, S., Bender, W., Bailey, J. M., Davidson, N., Nicolson, M. O., McAllister, R. M.: RD-114, baboon and woolly monkey viral RNAs compared in size and structure. Cell **7**, 609—620 (1976).

Kurth, R., Teich, N. M., Weiss, R., Oliver, R. T. D.: Natural human antibodies reactive with primate type-C viral antigens. Proc. nat. Acad. Sci. (Wash.) **74**, 1237—1241 (1977).

Lapin, B. A.: The epidemiologic and genetic aspects of an outbreak of leukemia among Hamadryas baboons of the Sukhumi monkey colony (Bibliotheca Haematologica, No. 39). (Dutcher, R., Chieco-Bianchi, L., eds.), 263—268. Basel: Karger 1973.

Larsen, C. J., Marty, M., Hamelin, R., Peries, J., Boiron, M., Tavitian, A: Search for nucleic acid sequences complementary to a murine oncornaviral genome in poly(A)-rich RNA of human leukemic cells. Proc. nat. Acad. Sci. (Wash.) **72**, 4900—4904 (1975).

Lau, A. F., Krzyzek, R. A., Brugge, J. S., Erikson, R. L., Schollmeyer, J., Faras, A. J.: Morphological revertants of an avian sarcoma virus-transformed mammalian cell line exhibit tumorigenicity and contain pp60 src. Proc. nat. Acad. Sci. (Wash.) **76**, 3904—3908 (1979).

Lavi, S., Winocour, E.: Acquisition of sequences homologous to host deoxyribonucleic acid by closed circular simian virus 40 deoxyribonucleic acid. J. Virol. **9**, 309—316 (1972).

Leong, J., Garapin, A., Jackson, N., Fanshier, L., Levinson, W., Bishop, J. M.: Virus-specific ribonucleic acid in cells producing Rous sarcoma virus: detection and characterization. J. Virol. **9**, 891—902 (1972).

Leuders, K. K., Kuff, E. L.: Sequences associated with intracisternal A particles are reiterated in the mouse genome. Cell **12**, 963—972 (1977).

Levy, J. A.: Xenotropic viruses: murine leukemia viruses associated with NIH Swiss, NZB and other mouse strains. Science **182**, 1151—1153 (1973).

Levy, S., Childs, G., Kedes, L.: Sea urchin nuclei use RNA polymerase II to transcribe discrete histone RNAs larger than messengers. Cell **15**, 151—162 (1978).

Lieber, M. M., Livingstone, D. M., Todaro, G. J.: Superinduction of endogenous type-C virus by 5-bromodeoxyuridine from transformed mouse clones. Science **181**, 443—444 (1973).

Lieber, M. M., Sherr, C. J., Todaro, G. J., Benveniste, R. E., Callahan, R., Coon, H. G.: Isolation from the Asian mouse *Mus caroli* of an endogenous type C virus related to infectious primate type C viruses. Proc. nat. Acad. Sci. (Wash.) **72**, 2315—2319 (1975).

Lin, F. H., Thormar, H.: Characterization of the ribonucleic acid from visna virus. J. Virol. **7**, 582—587 (1971).

Loni, M. C., Green, M.: Virus-specific DNA sequences in mouse and rat cells transformed by the Harvey and Moloney murine sarcoma viruses detected by in situ hybridization. Virology **63**, 40—47 (1975).

Lowy, D. R., Chattopadhyay, S. K., Teich, N. M., Rowe, W. P., Levine, A. S.: AKR murine leukemia virus genome: frequency of sequences in DNA of high-, low-, and non-virus-yielding mouse strains. Proc. nat. Acad. Sci. (Wash.) **71**, 3555—3559 (1974).

Lowy, D., Rowe, W., Teich, N., Hartley, J.: Murine leukemia virus high frequency activation *in vitro* by 5-iododeoxyuridine and 5-bromodeoxyuridine. Science **174**, 155—157 (1971).

Ludford, C. G., Purchase, H. G., Cox, H. W.: Duck infectious anemia virus associated with *Plasmodium lophurae*. Exp. Parasit. **31**, 29—38 (1972).

Lyons, M. J., Moore, D. H.: Isolation of the mouse mammary tumor virus: chemical and morphological studies. J. nat. Cancer Inst. **35**, 549—557 (1965).

Maio, J. J., Brown, F. L., Musich, P. R.: Subunit structure of chromatin and the organization of eukaryotic highly repetitive DNA: recurrent periodicities and models for the evolutionary origin of repetive DNA. J. molec. Biol. **177**, 637—655 (1977).

Maisel, J., Bender, W., Hu, S., Duesberg, P. H., Davidson, N.: Structure of the 50—70S RNA from Moloney sarcoma viruses. J. Virol. **25**, 384—394 (1978).

Maisel, J., Klement, V., Lai, M. M.-C., Ostertag, W., Duesberg, P.: Ribonucleic acid components of murine sarcoma and leukemia viruses. Proc. nat. Acad. Sci. (Wash.) **70**, 3536—3540 (1973).

Maisel, J., Scolnick, E., Duesberg, P.: Base sequence differences between the RNA components of Harvey sarcoma virus. J. Virol. **16**, 749—753 (1975).

Mak, T. W., Kurtz, S., Manaster, J., Housman, D.: Viral-related information in oncornavirus-like particles isolated from cultures of marrow cells from leukemic patients in relapse and remission. Proc. nat. Acad. Sci. (Wash.) **72**, 623—627 (1975).

Mak, T. W., Manaster, J., Howatson, A. F., McCulloch, E. A., Till, J. E.: Particles with characteristics of leukoviruses in cultures of marrow cells from leukemic patients in remission and relapse. Proc. nat. Acad. Sci. (Wash.) **71**, 4336 (1974).

Markham, P. D., Baluda, M. A.: Integrated state of oncornavirus DNA in normal chicken cells transformed by avian myeloblastosis virus. J. Virol. **12**, 721—732 (1973).

Marmur, J., Doty, P.: Heterogeneity in deoxyribonucleic acids. I. Dependence on composition of the configurational stability of deoxyribonucleic acids. Nature (Lond.) **183**, 1427—1429 (1959).

Mayer, R., Smith, R. G., Gallo, R. C.: Reverse transcriptase in normal rhesus monkey placenta. Science **185**, 864—867 (1974).

McAllister, R. M., Nicolson, M., Gardner, M. B., Rongey, R. W., Rasheed, S., Sarma, P. S., Huebner, R. J., Hatanaka, M., Orozlan, S., Gilden, R. V., Kabigting, A., Vernon, L. I.: C-type virus released from cultured human rhabdomyosarcoma cells. Nature (New Biol.) **235**, 3—6 (1972).

McGhee, R. B., Loftis, E.: A filterable proliferating factor stimulating autoimmunity in malarious and nonmalarious ducklings. Exp. Parasit. **22**, 299—308 (1968).

Melchior, W. B., von Hippel, P. H.: Alteration of the relative stability of dA · dT and dG · dC base pairs in DNA. Proc. nat. Acad. Sci. (Wash.) **70**, 298—302 (1973).

Mellors, R. C., Mellors, J. W.: Antigen related to mammalian type-C RNA viral p30 proteins is located in renal glomeruli in human systemic lupus erythematosus. Proc. nat. Acad. Sci. (Wash.) **73**, 233—237 (1976).

Melnick, J. L., Altenburg, B., Arnstein, P., Mirkovic, R., Tevethia, S. S.: Transformation of baboon cells with feline sarcoma virus. Intervirology **1**, 386—398 (1973).

Metzgar, R. S., Mohanakumar, T., Bolognesi, D. P.: Relationships between membrane antigens of human leukemic cells and oncogenic RNA virus structural components. J. exp. Med. **143**, 47—63 (1976).

Miller, N. R., Saxinger, W. C., Reitz, M. S., Gallagher, R. E., Wu, A. M., Gallo, R. C., Gillespie, D.: Systematics of RNA tumor viruses and virus-like particles of human origin. Proc. nat. Acad. Sci. (Wash.) **71**, 3177—3181 (1974).

Mintz, B., Illmensee, K.: Normal genetically mosaic mice produced from malignant teratocarcinoma cells. Proc. nat. Acad. Sci. (Wash.) **72**, 3585—3589 (1975).

Moloney, J.: Biological studies on a lymphoid-leukemia virus extracted from sarcoma 37. I. Origin and introductory investigations. J. nat. Cancer Inst. **24**, 933—947 1960).

MOLONEY, J. B.: A virus-induced rhabdomyosarcoma of mice. Nat. Cancer Inst. Monogr. **22**, 139—140 (1966).

MONDAL, H., GALLAGHER, R. E., GALLO, R. C.: RNA-directed DNA polymerase from human leukemic blood cells and from primate type C virus-producing cells: High and low molecular weight forms with variant biochemical and immunological properties. Proc. nat. Acad. Sci. (Wash.) **71**, 1194—1198 (1974).

MUKHERJEE, B. B., MOBRY, P. M.: Variations in hybridization of RNA from different mouse tissues and embryos to endogenous C-type virus DNA transcripts. J. gen. Virol. **28**, 129—135 (1975).

NAYAK, D. P.: Endogenous guinea pig virus: equability of virus specific DNA in normal, leukemic, and virus-producing cells. Proc. nat. Acad. Sci. (Wash.) **71**, 1164 to 1168 (1974).

NAYAK, D. P., DAVIS, A. R.: Endogenous oncornaviral DNA sequences: evidence for two classes of viral DNA sequences in guinea pig cells. J. Virol. **17**, 745—755 (1976).

NEIMAN, P.: Rous sarcoma virus nucleotide sequences in cellular DNA: measurement by RNA-DNA hybridization. Science **178**, 750—753 (1972).

NEIMAN, P. E.: Measurement of endogenous leukosis virus nucleotide sequences in the DNA of normal avian embryos by RNA-DNA hybridization. Virology **53**, 196—204 (1973).

NEIMAN, P.: Measurement of RD114 nucleotide sequences in feline cellular DNA. Nature (New Biol.) **244**, 62—64 (1973).

NEIMAN, P. E., PURCHASE, H. G., OKAZAKI, W.: Chicken leukosis virus genome sequences from normal chick cells and virus induced bursal lymphomas. Cell **4**, 311 to 319 (1975).

NEIMAN, P., WRIGHT, S. E., MCMILLAN, C., MACDONNEL, L.: Nucleotide sequence relationship of avian RNA tumor viruses: measurement of the deletion in a transformation-defective mutant of Rous sarcoma virus. J. Virol. **13**, 837—846 (1974).

NIMAN, H. L., GARDNER, M. B., STEPHENSON, J. R., ROY-BURMAN, P.: Endogenous RD114 virus genome expression in malignant tissues of domestic cats. J. Virol. **23**, 578—586 (1977).

NOOTER, K., AARSSEN, A. M., BENTVELZEN, P., D'GROOT, F. G.: Isolation of an infectious C type oncornavirus from human leukemic bone marrow cells. Nature (Lond.) **256**, 595—597 (1975).

NOOTER, K., BENTVELZEN, P., ZURCHER, C., RHIM, J.: Detection of human C-type helper viruses in human leukemic bone marrow with murine sarcoma virus-transformed human and rat nonproducer cells. Int. J. Cancer **19**, 59—65 (1977).

NOOTER, K., OVERDEVEST, J., DUBBES, R., KOCH, G., BENTVELZEN, P., ZURCHER, C., COOLEN, J., CALAFAT, J.: Type-C oncornavirus isolate from human leukemic bone marrow: Further in vitro and in vivo characterization. Int. J. Cancer **21**, 27—34 (1978).

OFFICER, J. E., TECSON, N., ESTES, J. D., FONTANILLA, R., RONGEY, W., GARDNER, M. B.: Isolation of a neurotropic type-C virus. Science **181**, 945—947 (1973).

OKABE, H., GILDEN, R. V., HATANAKA, M.: Extensive homology of RD114 virus DNA with RNA of feline origin. Nature (New Biol.) **244**, 54—56 (1973).

OKABE, H., GILDEN, R. V., HATANAKA, M.: Specificity of the DNA product of RNA-dependent DNA polymerase in type-C viruses: a quantitative analysis. Proc. nat. Acad. Sci. (Wash.) **70**, 3923—3927 (1973).

OKABE, H., TWIDDY, E., GILDEN, R. V., HATANAKA, M., HOOVER, E. A., OLSEN, R. G.: FeLv-related sequences in DNA from a FeLV-free cat colony. Virology **69**, 798 to 801 (1976).

PANEM, S., PROCHOWNIK, E., KNISH, W. H., KIRSTEN, W. H.: Cell generation and type-C virus expression in the human embryonic cell strain HEL-12. J. gen. Virol. **35**, 487—495 (1977).

PANEM, S., PROCHOWNIK, E. V., REALE, R. R., KIRSTEN, W. H.: Isolation of C type virions from a normal human fibroblast strain. Science **189**, 297—299 (1975).

PANG, R. H. L., PHILLIPS, L. A., HAAPALA, D. K.: Characterization of Gazdar murine sarcoma virus by nucleic acid hybridization and analysis of viral expression in cells. J. Virol. **24**, 551—556 (1977).

PARAN, M., GALLO, R. C., RICHARDSON, L. S. WU, A. M.: Adrenal corticosteroids enhance production of type-C virus induced by 5-iododeoxyuridine from cultured mouse fribroblasts. Proc. nat. Acad. Sci. (Wash.) **70**, 2391—2395 (1973).

PARKS, W. P., SCOLNICK, E. M.: *In vitro* translation of Harvey murine sarcoma virus RNA. J.Virol. **22**, 711—719 (1977).

PAYNE, L. N., CHUBB, R. C.: Studies on the nature and genetic control of an antigen in normal chick embryos which reacts in the COFAL test. J. gen. Virol. **3**, 379—391 (1968).

PEEBLES, P. T., SCOLNICK, E. M., HOWK, R. S.: Increased sarcoma virus RNA in cells transformed by leukemia viruses: model for leukemogenesis. Science **192**, 1143 to 1145 (1976).

PINCUS, T., HARTLEY, J. W., ROWE, W. P.: A major genetic locus affecting resistance to infection with murine leukemia viruses. I. Tissue culture studies of naturally occurring viruses. J. exp. Med. **133**, 1219—1233 (1971).

PINCUS, T., ROWE, W. P., LILLY, F.: A major genetic locus affecting resistance to infection with murine leukemia viruses. II. Apparent identity to a major locus described for resistance to Friend murine leukemia virus. J. exp. Med. **133**, 1234 to 1241 (1971).

PIPER, C. T., ABT, D. A., FERRER, J. F., MARSHAK, R. R.: Seroepidemiological evidence for horizontal transmission of bovine type-C virus. Cancer Res. **35**, 2714—2716 (1975).

POSTE, G., FLOOD, M. K.: Cells transformed by temperature-sensitive mutants of avian sarcoma virus cause tumors *in vivo* at permissive and nonpermissive temperature. Cell **17**, 789—900 (1979).

PRIORI, E. S., DMOCHOWSKI, L., MYERS, B., WILBUR, J. R.: Constant production of type-C virus particles in a continuous tissue culture derived from pleural effusion cells of a lymphoma patient. Nature (New Biol.) **232**, 61—62 (1973).

PROCHOWNIK, E., KIRSTEN, W.: Inhibition of reverse transcriptases of primate type-C viruses by 7S immunoglobin form patients with leukemia. Nature (Lond.) **260**, 64—66 (1976).

PROCHOWNIK, E., PANEM, S., KIRSTEN, W.: Type-C virus induced by iododeoxyuridine in the human embryonic cell strain HEL-12. J. gen. Virol. **42**, 399—403 (1979).

PURCHIO, A. F., ERIKSON, E., BRUGGE, J. S., ERIKSON, R. L.: Identification of a polypeptide encoded by the avian sarcoma virus src gene. Proc. nat. Acad. Sci. (Wash.) **75**, 1567—1571 (1978).

RASHEED, S., GARDNER, M. B., CHAN, E.: Amphotropic host range of naturally-occurring wild mouse leukemia viruses. J. Virol. **19**, 13—18 (1976).

RAUSCHER, F. J.: A virus-induced disease of mice characterized by erythrocytopoesis and lymphoid leukemia. J. nat. Cancer Inst. **29**, 515—543 (1962).

REITZ, M. S., MILLER, N. R., WONG-STAAL, F., GALLAGHER, R. E., GALLO, R. C., GILLESPIE, D. H.: Primate type C virus nucleic acid sequences (woolly monkey and baboon types) in tissues from a patient with acute myelogenous leukemia and in viruses isolated from cultured cells of the same patient. Proc. nat. Acad. Sci. (Wash.) **73**, 2113—2117 (1976).

RICE, N. R., STRAUSS, N. A.: Relatedness of mouse sattelite deoxyribonucleic acid to deoxyribonucleic acid of various Mus species. Proc. nat. Acad. Sci. (Wash.) **70**, 3546—3550 (1973).

RINGOLD, G. M., SHANK, P. R., YAMAMOTO, K. R.: Production of unintegrated mouse mammary tumor virus DNA in infected rat hepatoma cells is a secondary action of dexamethasone. J. Virol. **26**, 93—101 (1978).

RISSER, R., STOCKERT, E., OLD, L.: Abelson antigen: A viral tumor antigen that is also a differentiation antigen of BALB/c mice. Proc. nat. Acad. Sci. (Wash.) **75**, 3918 to 3922 (1978).

RITOSSA, F.: Unstable redundancy of genes for ribosomal RNA. Proc. nat. Acad. Sci. (Wash.) **60**, 509—513 (1968).

ROBINSON, H. L.: Isolation of noninfectious particles containing Rous sarcoma virus RNA from the medium of Rous sarcoma virus-transformed nonproducer cells. Proc. nat. Acad. Sci. (Wash.) **57**, 1655—1662 (1967).

ROBINSON, W. S., BALUDA, M. A.: The nucleic acid from avian myeloblastosis virus compared with the RNA from the Bryan strain of Rous sarcoma virus. Proc. nat. Acad. Sci. (Wash.) **54**, 1686—1692 (1965).

ROBINSON, W. S., PITKANEN, A., RUBIN, H.: The nucleic acid of the Bryan strain of Rous sarcoma virus: purification of the virus and isolation of the nucleic acid. Proc. nat. Acad. Sci. (Wash.) **54**, 137—144 (1965).

ROLFE, R., MESELSON, M.: The relative homogeneity of microbial DNA. Proc. nat. Acad. Sci. (Wash.) **45**, 1039—1043 (1959).

ROSENBERG, H., SINGER, M., ROSENBERG, M.: Highly reiterated sequences of SIMIAN-SIMIANSIMIANSIMIAMSIMIAN. Science **200**, 394—402 (1978).

ROSENBERG, M., SEGAL, S., KUFF, E., SINGER, M.: The nucleotide sequence of repetitive monkey DNA found in defective simian virus 40. Cell **11**, 845—857 (1977).

ROWE, W. P.: Studies of genetic transmission of murine leukemia virus by AKR mice. J. exp. Med. **136**, 1272—1285 (1972).

ROY-BURMAN, P., KAPLAN, M. B.: Nucleotide composition of the RNA from RD-114 virions. Biochem. biophys. Res. Commun. **48**, 1354—1361 (1972).

ROY-BURMAN, P., KLEMENT, V.: Derivation of mouse sarcoma virus (Kirsten) by acquisition of genes from heterologous host. J. gen. Virol. **28**, 193—198 (1975).

RUBSAMEN, H., FRIIS, R. R., BAUER, H.: Src gene product from different strains of avian sarcoma virus: Kinetics and possible mechanism of heat inactivation of protein kinase activity from cells infected by transformation-defective, temperature-sensitive mutant and wild-type virus. Proc. nat. Acad. Sci. (Wash.) **76**, 967—971 (1979).

RUPRECHT, R. M., GOODMAN, N. C., SPIEGELMAN, S.: Determination of natural host taxonomy of RNA tumor viruses by molecular hybridization: application to RD114, a candidate human virus. Proc. nat. Acad. Sci. (Wash.) **70**, 1437—1441 (1973).

SABRAN, J., HSU, T., YEATER, C., KAJI, A., MASON, W., TAYLOR, J.: Analysis of integrated avian RNA tumor virus DNA in transformed chicken, duck and quail fibroblasts. J. Virol. **29**, 170—178 (1979).

SARIN, P. S., GALLO, R. C.: RNA directed DNA polymerase. In: Int. Rev. Science (BURTON, K., ed.), Vol. 6, chapter 8. Butterworth 1974

SARNGADHARAN, M. G., SARIN, P. S., REITZ, M. S., GALLO, R. C.: Reverse transcriptase activity of human leukemic cells: Purification of the enzyme, response to AMV 70S RNA, and characterization of the DNA product. Nature (New Biol.) **240**, 67 (1972).

SCHIDLOVSKY, G., AHMED, M.: C-type virus particles in placentas and fetal tissues of rhesus monkeys. J. nat. Cancer Inst. **51**, 225—233 (1973).

SCHINCAROL, A. L., JOKLIK, W. K.: Early synthesis of virus-specific RNA and DNA in cells rapidly transformed with Rous sarcoma virus. Virology **56**, 532—548 (1973).

SCHOOLMAN, H. M., SPURRIER, W., SCHWARTZ, S. O., SZANTO, P. B.: Studies in leukemia III. The induction of leukemia in Swiss mice by means of cell-free filtrates of leukemia mouse brain. Blood **12**, 694—700 (1957).

SCHWARTZ, D. E., ZAMECNIK, P. C., WEITH, H. L.: Rous sarcoma virus genome is terminally redundant: the 3' sequence. Proc. nat. Acad. Sci. (Wash.) **74**, 994—998 (1977).

SCOLNICK, E. M., GOLDBERG, R. J., WILLIAMS, D.: Characterization of rat genetic sequences in Kirsten sarcoma virus: distinct class of rat endogenous type C viral sequences. J. Virol. **18**, 559—566 (1976).

SCOLNICK, E. M., HOWK, R. S., ANISOWICZ, A., PEEBLES, P. T., SCHER, D. D., PARKS, W. P.: Separation of sarcoma virus-specific and leukemia virus-specific genetic sequences of Moloney sarcoma virus. Proc. nat. Acad. Sci. (Wash.) **72**, 4650—4654 (1975).

SCOLNICK, E. M., MARYAK, J. M., PARKS, W. P.: Levels of rat cellular RNA homologous
  to either Kirsten sarcoma virus or to rat type-C virus in cell lines derived from
  Osborne-Mendel rats. J. Virol. **14**, 1435—1444 (1974).
SCOLNICK, E. M., PAPAGEORGE, A. G., SHIH, T. Y.: Guanine nucleotide-binding
  activity as an assay for src protein of rat-derived murine sarcoma viruses. Proc.
  nat. Acad. Sci. (Wash.) **76**, 5355—5359 (1979).
SCOLNICK, E. M., PARKS, W. P.: Harvey sarcoma virus: a second murine type-C
  sarcoma virus with rat genetic information. J. Virol. **13**, 1211—1219 (1974).
SCOLNICK, E. M., PARKS, W., KAWAKAMI, T., KOHNE, D., OKABE, H., GILDEN, R. V.,
  HATANAKA, M.: Primate and murine type-C viral nucleic acid association kinetics:
  analysis of model systems and natural tissues. J. Virol. **13**, 363—369 (1974).
SCOLNICK, E. M., RANDS, E., WILLIAMS, D., PARKS, W. P.: Studies on the nucleic acid
  sequences of Kirsten sarcoma virus: a model for formation of a mammalian RNA-
  containing sarcoma virus. J. Virol. **12**, 458—463 (1973).
SCOLNICK, E. M., WILLIAMS, D., MARYAK, J., VARS, W., GOLDBERG, R. J., PARKS,
  W. P.: Type C particle positive and type C particle negative rat cell lines: characterization of the coding capacity of endogenous sarcoma virus-specific RNA. J. Virol.
  20, 570—582 (1976).
SEEMAN, N. C., ROSENBERG, J. M., RICH, A.: Sequence specific recognition of double
  helical nucleic acids by proteins. Proc. nat. Acad. Sci. (Wash.) **73**, 804—808
  (1976).
SHANK, P. R., VARMUS, H. E.: Virus-specific DNA in the cytoplasm of avian sarcoma
  virus-infected cells is a precursor of covalently closed circular viral DNA in the
  nucleus. J. Virol. **25**, 104—114 (1978).
SHEARER, R., SMUCKLER, E. A.: A search for gene derepression in RNA of primary rat
  hepatomas. Cancer Res. **31**, 2104—2109 (1971).
SHEARER, R. W., SMUCKLER, E. A.: Altered regulation of the transport of RNA from
  nucleus to cytoplasm in rat hepatoma cells. Cancer Res. **32**, 339—342 (1972).
SHERR, C. J., BENVENISTE, R. E., TODARO, G. J.: Endogenous mink *(mustella vison)*
  type-C virus isolated from sarcoma virus-transformed mink cells. J. Virol. **25**,
  738—749 (1978).
SHERR, C. J., TODARO, G. J.: Type C virus antigens in man. Antigens related to endogenous primate virus in human tumors. Proc. nat. Acad. Sci. (Wash.) **71**, 4703—
  4707 (1974).
SHERR, C. J., TODARO, G. J.: Primate type C virus p 30 antigen in cells from humans
  with acute leukemia. Science **187**, 855—857 (1975).
SHERWIN, S. A., RAPP, U. R., BENVENISTE, R. E., SEN, A., TODARO, G. J.: Rescue of
  endogenous 30S retroviral sequences from mouse cells by baboon type C virus.
  J. Virol. **26**, 257—264 (1978).
SHERWIN, S., SLISKI, A., TODARO, G.: Human melanoma cells have both nerve growth
  factor and nerve growth factor-specific receptors on their cell surfaces. Proc. nat.
  Acad. Sci. (Wash.) **76**, 1288—1292 (1979).
SHIH, R. Y., WEEKS, M. O., YOUNG, H. A., SCOLNICK, E. M.: P21 of Kirsten murine
  sarcoma virus is thermolabile in a viral mutant temperature-sensitive for the
  maintenance of transformation. J. Virol. **31**, 546—556 (1979).
III, I. Y., YOUNG, H. A., COFFIN, J. M., SCOLNICK, E. M.: Physical map of Kirsten
  sarcoma virus genome as determined by fingerprinting T1-resistant oligonucleotides. J. Virol. **25**, 238—253 (1978).
HOYAB, M., BALUDA, M. A.: Acquisition of viral DNA sequences in target organs of
  chickens infected with avian myeloblastosis virus. J. Virol. **16**, 783—789 (1975).
OYAB, M., BALUDA, M. A.: Homology between avian oncornavirus RNA and DNA
  from several avian species. J. Virol. **16**, 1492—1502 (1975).
HOYAB, M., BALUDA, M.: Ribonucleotide sequence homology among avian oncorna
  viruses. J. Virol. **17**, 106—113 (1976).
SHOYAB, M., BALUDA, M. A., EVANS, R.: Acquisition of new DNA sequences after
  infection of chicken cells with avian myeloblastosis virus. J. Virol. **13**, 331—339
  (1974).

SHOYAB, M., DASTOOR, M. N., BALUDA, M. A.: Evidence for tandem integration of avian myeloblastosis virus DNA with endogenous proviruses in leukemic chicken cells. Proc. nat. Acad. Sci. (Wash.) **73**, 1749—1753 (1976).

SHOYAB, M., EVANS, R., BALUDA, M. A.: Presence in leukemic cells of avian myeloblastosis-specific DNA sequences absent in normal chicken cells. J. Virol. **14**, 47—49 (1974).

SHOYAB, M., MARKHAM, P. D., BALUDA, M. A.: Host induced alternation of avian sarcoma virus B 77 genome. Proc. nat. Acad. Sci. (Wash.) **72**, 1030—1035 (1975).

SMITH, G. P.: Unequal crossover and the evolution of multigene families. Cold Spr. Harb. Symp. quant. Biol. **38**, 507—513 (1973).

SMITH, G. P.: Evolution of repeated DNA sequences by unequal crossover. Science **191**, 528—535 (1976).

SMITH, R. E., DAVIDS, L. J., NEIMAN, P. E.: Comparison of an avian osteopetrosis virus with an avian lymphomatosis virus by RNA-DNA hybridization. J. Virol. **17**, 160—167 (1976).

SMITH, R. E., MUSCOVICI, C.: The oncogenic effects of nontransforming viruses from avian myeloblastosis virus. Cancer Res. **29**, 1356—1366 (1969).

SMUCKLER, E. A., KOPLITZ, M.: Altered nuclear RNA transport associated with carcinogen intoxication in rats. Biochem. biophys. Res. Commun. **55**, 499—507 (1973).

SNYDER, H. W., PINCUS, T., FLEISSNER, E.: Specificities of human immunoglobulins reactive with antigens in preparations of several mammalian type-C viruses. Virology **75**, 60—73 (1976).

STEELE, L. K., LAUBE, H., CHANDRA, P.: Biochemical and serological characteristics of reverse transcriptase from human spleen in a case of childhood myelofibrotic syndrome. Cancer Letters **2**, 291—297 (1977).

STEHELIN, D., VARMUS, H. E., BISHOP, J. M., VOGT, P. K.: DNA related to the transforming genes of avian sarcoma viruses is present in normal avian DNA. Nature (Lond.) **260**, 170—173 (1976).

STEPHENSON, J. R., GREENBERGER, J. S., AARONSON, S. A.: Oncogenicity of an endogenous C-type virus chemically activated from mouse cells in culture. J. Virol. **13**, 237—240 (1974).

STOCKERT, E., OLD, L. J., BOYSE, E. A.: The G_{IX} system. J. exp. Med. **133**, 1334 to 1355 (1971).

STRAND, M., AUGUST, J. T.: Type-C RNA virus gene expression in human tissue. J. Virol. **14**, 1584—1596 (1974).

SUEOKA, N., MARMUR, J., DOTY, P.: Heterogeneity in deoxyribonucleic acids. Nature (Lond.) **183**, 1429—1431 (1959).

SVEDA, M. M., SOEIRO, R.: Host restriction of Friend leukemia virus: Synthesis and integration of the provirus. Proc. nat. Acad. Sci. (Wash.) **73**, 2356—2360 (1976).

SWEET, R. W., GOODMAN, N. C., CHO, J. R., RUPRECHT, R. M., REDFIELD, R. R., SPIEGELMAN, S.: The presence of unique DNA sequences after viral induction of leukemia in mice. Proc. nat. Acad. Sci. (Wash.) **71**, 1705—1709 (1974).

TAVITIAN, A., LARSEN, C. J., HAMELIN, R., BOIRON, M.: Murine and simian C-type viruses: Sequences detected in the RNA of human leukemic cells by the c-DNA probes. In: Modern Trends in Human Leukemia II (NETH, R., GALLO, R. C., MANNWEILER, K., MOLONEY, W. C., eds.), 451—455. München: J. F. Lehmann 1976.

TAYLOR, J., ILLMENSEE, R.: Site on the RNA of an avian sarcoma virus at which primer is bound. J. Virol. **16**, 553—558 (1975).

TEICH, N. M., Weiss, R. A., SALAHUDDIN, S. Z., GALLAGHER, R. E., GILLESPIE, D. H., GALLO, R. C.: Infective transmission and characterization of a C type virus released by cultured human myeloid leukemia cells. Nature (Lond.) **256**, 551—555 (1975).

TEMIN, H. M.: Homology between RNA from Rous sarcoma virus and DNA from Rous sarcoma virus-infected cells. Proc. nat. Acad. Sci. (Wash.) **52**, 323—327 (1964).

TEMIN, H. M.: The protovirus hypothesis: speculations on the significance of RNA-directed DNA synthesis for normal development and for carcinogenesis. J. nat. Cancer Inst. **46**, III—VII (1971).

TEMIN, H. M.: The cellular and molecular biology of RNA tumor viruses especially avian leukosis-sarcoma and their relatives. Advanc. Cancer Res. **19**, 47—104 (1974).

TEMIN, H. M., BALTIMORE, D.: RNA-directed DNA synthesis and RNA tumor viruses. Advanc. Virus Res. **17**, 129—186 (1972).

TEMIN, H. M., MITZUTANI, H.: RNA-dependent DNA polymerase in virions of Rous sarcoma virus. Nature (Lond.) **226**, 1211—1213 (1970).

TEREBA, A., SKOOG, L., VOGT, P.: RNA tumor virus sequences in nuclear DNA of several avian species. Virology **65**, 524—534 (1975).

THEILEN, G. H., ZEIGEL, R. F., TWIEHAUS, M. J.: Biological studies with RE virus (strain T) that induces reticuloendotheliosis in turkeys, chickens and Japanese quail. J. nat. Cancer Inst. **37**, 731—743 (1966).

THEILEN, G., GOULD, D., FOWLER, M., DUNGWORTH, D.: C type virus in tumor tissue of a woolly monkey (lagothrix) with fibrosarcoma. J. nat. Cancer Inst. **47**, 881—884 (1971).

TILGHMAN, S. M., TIEMEIER, D. C., SEIDMAN, J. G., PETERLIN, B. M., SULLIVAN, M., MAIZEL, J. V., LEDER, P.: Intervening sequence of DNA identified in the structural portion of a mouse β-globin gene. Proc. nat. Acad. Sci. (Wash.) **75**, 725—729 (1978).

TODARO, G. J., BENVENISTE, R. E., SHERR, C.: Interspecies transfer of RNA tumor virus genes: implications for the search for "human" type-C viruses. ICN-UCLA Symp. molec. cell. Biol. **4**, 385—408 (1976).

TODARO, G. J., GALLO, R. C.: Immunological relationship of DNA polymerase from human acute leukemia cells and primate and mouse leukemia virus reverse transcriptase. Nature (Lond.) **244**, 206—209 (1973).

TODARO, G. J., GALLO, R. C.: Oncogenic RNA viruses. Semin. Oncol. **3**, 81—96 (1976).

TODARO, G. J., HUEBNER, R. J.: The viral oncogene hypothesis: new evidence. Proc. nat. Acad. Sci. (Wash.) **69**, 1009—1015 (1972).

TODARO, G. J., LIEBER, M. M., BENVENISTE, R. E., SHERR, C. J., GIBBS, C. J., GAJDUSEK, D. C.: Infectious primate type C viruses: Three isolates belonging to a new subgroup from the brains of normal gibbons. Virology **67**, 335—343 (1975).

TODARO, G. J., SHERR, C. J., BENVENISTE, R. E., LIEBER, M. M., MELNICK, J. L.: Type-C viruses of baboons: isolation from normal cell cultures. Cell **2**, 55—61 (1974).

TOOZE, J.: The Molecular Biology of Tumor Viruses. Cold Spring Harbor Monograph (1973).

TRAGER, W.: A new virus of chicks interfering with the development of malarial parasite *(Plasmodium lophurae)*. Proc. Soc. exp. Biol. (N. Y.) **101**, 578—582 (1959).

TRONICK, S. R., CABRADILLO, C. D., AARONSON, S. A., HASELTINE, W. A.: 5'-terminal nucleotide sequences of mammalian type-C helper viruses are conserved in the genomes of replication-defective mammalian transforming viruses. J. Virol. **26**, 570—576 (1978).

TROXLER, D. H., BOYARS, J. K., PARKS, W. P., SCOLNICK, E. M.: Friend strain of focus-forming virus: a recombinant between mouse type C ecotropic viral sequences and sequences related to xenotropic viruses. J. Virol. **22**, 361—372 (1977).

TROXLER, D. H., LOWY, D., HOWK, R., YOUND, H., SCOLNICK, E. M.: The spleen focus-forming virus is a recombinant between ecotropic murine type-C virus, and the *env* gene region of xenotropic type-C virus. Proc. nat. Acad. Sci. (Wash.) **74**, 4671—4675 (1977).

TSUCHIDA, N., GREEN, M.: Intracellular viral RNA species in mouse cells nonproductively transformed by the murine sarcoma virus. J. Virol. **14**, 587—591 (1974).

TWIEHAUS, M. J., ROBINSON, F. R.: (Referred to as unpublished data in THEILEN, ZEIGEL, and TWIEHAUS, 1966).

VARMUS, H. E., SHANK, P. R.: Unintegrated viral DNA is synthesized in the cytoplasm of avian sarcoma virus-transformed duck cells by viral DNA polymerase. J. Virol. **18**, 567—573 (1976).

VARMUS, H. E., BISHOP, J. M., VOGT, P. K.: Appearance of virus-specific DNA in mammalian cells following transformation by Rous sarcoma virus. J. molec. Biol. **74**, 613—626 (1973).

VARMUS, H. E., WEISS, R. A., FRIIS, R. R., LEVINSON, W., BISHOP, J. M.: Detection of avian tumor virus-specific nucleotide sequences in avian cell DNAs. Proc. nat. Acad. Sci. (Wash.) **69**, 20—24 (1972).

VARMUS, H. E., HEASLEY, S., KUNG, H.-J., OPPERMAN, H., SMITH, V. C., BISHOP, J. M., SHANK, P. R.: Kinetics of synthesis, structure and purification of avian sarcoma virus-specific DNA made in the cytoplasm of acutely infected cells. J. molec. Biol. **120**, 55—82 (1978).

VIOLA, M. V., FRAZIER, M., WIERNIK, P. H., McCREDIE, K. B., SPIEGELMAN, S.: Reverse transcriptase in leukocytes of leukemic patients in remission. New Engl. J. Med. **294**, 75—80 (1976).

VOGELSTEIN, B., GILLESPIE, D.: Preparative and analytical removal of DNA from agarose. Proc. nat. Acad. Sci. (Wash.) **76**, 615—619 (1979).

VOGT, P. K., FRIIS, R.: An avian leukosis virus related to RSV (o): Properties and evidence for helper activity. Virology **43**, 223—234 (1971).

VOSIKA, G. J., KRIVIT, W., GERRARD, J. M., COCCIA, P. F., NESBIT, M. E., COALSON, J. J., KENNEDY, B. J.: Oncornavirus-like particles from cultured bone marrow cells preceding leukemia and malignant histiocytosis. Proc. nat. Acad. Sci. (Wash.) **72**, 2804—2808 (1975).

WADA, A., TACHIBANA, H., GOTOH, O., TAKANAMI, M.: Long range homogeneity of physical stability in double-stranded DNA. Nature (Lond.) **263**, 439—440 (1976).

WANG, L. H., DUESBERG, P. H., BEEMAN, K., VOGT, P.: Mapping T1-resistant oligonucleotides, etc. J. Virol. **16**, 1050—1070 (1975).

WARING, M., BRITTEN, R. J.: Nucleotide sequence repitition: A rapidly reassociating fraction of mouse DNA. Science **154**, 791—794 (1966).

WEISS, R. A.: The host range of Bryan strain Rous sarcoma virus synthesized in the absence of helper virus. J. gen. Virol. **5**, 511—528 (1969).

WEISS, R. A., FRIIS, R. R., KATZ, E., VOGT, P.: Induction of avian tumor viruses in normal cells by physical and chemical carcinogens. Virology **46**, 920—938 (1971).

WEISS, R. A., MASON, W. S., VOGT, P. K.: Genetic recombinants and heterozygotes derived from endogenous and exogenous avian RNA tumor viruses. Virology **52**, 535—552 (1973).

WETMUR, J. G., DAVIDSON, N.: Kinetics of renaturation of DNA. J. molec. Biol. **31**, 349—370 (1968).

WOLFE, L., DEINHARDT, F., THEILEN, G., KAWAKAMI, T., BUSTAD, L.: Induction of tumors in marmoset monkeys by simian sarcoma virus type I (Lagothrix): A preliminary report. J. nat. Cancer Inst. **48**, 1905 (1972).

WOLFE, L. G., DEINHARDT, F., THEILEN, G. H., RABIN, H., KAWAKAMI, T., BUSTAD, L. K.: Induction of tumors in marmosets by simian sarcoma virus, type 1 (lagothrix): a preliminary report. J. nat. Cancer Inst. **47**, 1115—1120 (1971).

WONG-STAAL, F., GALLO, R. C., GILLESPIE, D.: Genetic relationship of a primate RNA tumor virus genome to genes in normal mice. Nature (Lond.) **256**, 670—672 (1975).

WONG-STAAL, F., GILLESPIE, D., GALLO, R. C.: Proviral sequences of baboon endogenous virus in DNA of leukemic tissue from seven patients with myelogenous leukemia. Nature (Lond.) **262**, 190—194 (1976).

WONG-STAAL, F., REITZ, M. S., TRAINOR, C. D., GALLO, R. C.: Murine intracisternal type A particles: a biochemical characterization. J. Virol. **16**, 887—896 (1975).

WU, A. M., REITZ, M. S., PARAN, M., GALLO, R. C.: On the mechanism of stimulation of murine type-C RNA tumor virus production by gluticosteroids: posttranscriptional effects. J. Virol. **14**, 802—811 (1974).

WU, A, M., TING, R. C., PARAN, M., GALLO, R. C.: Cordycepin inhibits induction of murine leukovirus production by 5-iododeoxyuridine. Proc. nat. Acad. Sci. (Wash.) **69**, 3820—3824 (1972).

YANG, S. S., WIVEL, N. A.: Analysis of high molecular weight ribonucleic acid associated with intracisternal A particles. J. Virol. **11**, 287—298 (1973).

YOSHIMURA, F. K., WEINBERG, R. A.: Restriction endonuclease cleavage of linear and closed circular murine leukemia viral DNAs: Discovery of a smaller circular form. Cell **16**, 323—332 (1979).

VIROLOGY MONOGRAPHS

Volume 1:

ECHO Viruses
By **H. A. Wenner** and **A. M. Behbehani**

Reoviruses
By **L. Rosen**

4 figures. IV, 107 pages. 1968.

ISBN 3-211-80889-2 (Wien)
ISBN 0-387-80889-2 (New York)

Volume 2:

The Simian Viruses
By **R. N. Hull**

Rhinoviruses
By **D. A. J. Tyrrell**

19 figures. IV, 124 pages. 1968.

ISBN 3-211-80890-6 (Wien)
ISBN 0-387-80890-6 (New York)

Volume 3:

Cytomegaloviruses
By **J. B. Hanshaw**

Rinderpest Virus
By **W. Plowright**

Lumpy Skin Disease Virus
By **K. E. Weiss**

26 figures. IV, 131 pages. 1968.

ISBN 3-211-80891-4 (Wien)
ISBN 0-387-80891-4 (New York)

Volume 4:

The Influenza Viruses
By **L. Hoyle**

58 figures. IV, 375 pages. 1968.

ISBN 3-211-80892-2 (Wien)
ISBN 0-387-80892-2 (New York)

Volume 5:

Herpes Simplex and Pseudorabies Viruses
By **A. S. Kaplan**

14 figures. IV, 115 pages. 1969.

ISBN 3-211-80932-5 (Wien)
ISBN 0-387-80932-5 (New York)

Volume 6:

Interferon
By **J. Vilček**

4 figures. IV, 141 pages. 1969.

ISBN 3-211-80933-3 (Wien)
ISBN 0-387-80933-3 (New York)

Volume 7:

Polyoma Virus
By **B. E. Eddy**

Rubella Virus
By **E. Norrby**

22 figures. IV, 174 pages. 1969.

ISBN 3-211-80934-1 (Wien)
ISBN 0-387-80934-1 (New York)

VIROLOGY MONOGRAPHS continued

Volume 8:

Spontaneous and Virus Induced Transformation in Cell Culture

By **J. Pontén**

35 figures. IV, 253 pages. 1971.

ISBN 3-211-80991-0 (Wien)
ISBN 0-387-80991-0 (New York)

Volume 9:

African Swine Fever Virus

By **W. R. Hess**

Bluetongue Virus

By **P. G. Howell** and
D. W. Verwoerd

5 figures. IV, 74 pages. 1971.

ISBN 3-211-81006-4 (Wien)
ISBN 0-387-81006-4 (New York)

Volume 10:

Lymphocytic Choriomeningitis Virus

By **F. Lehmann-Grube**

16 figures. V, 173 pages. 1971.

ISBN 3-211-81017-X (Wien)
ISBN 0-387-81017-X (New York)

Volume 11:

Canine Distemper Virus
By **M. J. G. Appel** and
J. H. Gillespie

Marburg Virus
By **R. Siegert**
50 figures. IV, 153 pages. 1972.
ISBN 3-211-81059-5 (Wien)
ISBN 0-387-81059-5 (New York)

Volume 12:

Varicella Virus
By **D. Taylor-Robinson** and
A. E. Caunt
10 figures. IV, 88 pages. 1972.
ISBN 3-211-81065-X (Wien)
ISBN 0-387-81065-X (New York)

Volume 13:

Lactic Dehydrogenase Virus
By **K. E. K. Rowson** and
B. W. J. Mahy
54 figures. IV, 121 pages. 1975.
ISBN 3-211-81270-9 (Wien)
ISBN 0-387-81270-9 (New York)

Volume 14:

Molecular Biology of Adenoviruses
By **L. Philipson, U. Pettersson,**
and **U. Lindberg**
20 figures. IV, 115 pages. 1975.
ISBN 3-211-81284-9 (Wien)
ISBN 0-387-81284-9 (New York)

Volume 15:

The Parvoviruses
By **G. Siegl**

1 figure. IV, 109 pages. 1976.

ISBN 3-211-81355-1 (Wien)
ISBN 0-387-81355-1 (New York)

Volume 16:

Dengue Viruses
By **R. W. Schlesinger**

34 figures. IV, 132 pages. 1977.

ISBN 3-211-81406-X (Wien)
ISBN 0-387-81406-X (New York)

Date Due